MAKING THE
INVISIBLE
VISIBLE

A History of the Spitzer Infrared Telescope Facility (1971–2003)

MONOGRAPHS IN AEROSPACE HISTORY, NO. 47

Renee M. Rottner

National Aeronautics and Space Administration

Office of Communications
NASA History Division
Washington, DC 20546

NASA SP-2017-4547

ON THE COVER
Front: Giant star Zeta Ophiuchi and its effects on the surrounding dust clouds
Back (top left to bottom right): Orion, the Whirlpool Galaxy, galaxy NGC 1292, RCW 49 nebula, the center of the Milky Way Galaxy, "yellow balls" in the W33 Star forming region, Helix Nebula, spiral galaxy NGC 2841

Contents

    v Acknowledgments

  vii List of Figures and Tables

   xi Introduction

1 CHAPTER 1
Ancient Views

13 CHAPTER 2
Getting Infrared Astronomy Off the Ground

29 CHAPTER 3
Making the Case for SIRTF

41 CHAPTER 4
SIRTF as a Shuttle-Based Infrared Telescope

61 CHAPTER 5
Selling It

83 CHAPTER 6
Out of Step

101 CHAPTER 7
From Orphan to Poster Child

117 CHAPTER 8
Constructing SIRTF

137 CHAPTER 9
Making the Invisible Visible

153 APPENDIX A
Contributors to SIRTF

159 Bibliography

173 The NASA History Series

183 Index

Acknowledgments

Over a thousand people helped make the Space Infrared Telescope Facility (SIRTF) a reality. Dozens more helped make this monograph a reality. But one person stands out—neither this monograph nor SIRTF would have been possible without the unflagging support of Mike Werner (Jet Propulsion Laboratory)—I dedicate this monograph to him.

I would like to express my gratitude to the people who gave of their time to be interviewed for this history: Fred Witteborn (Ames) imagined SIRTF before the technology existed, while Giovanni Fazio (Harvard-Smithsonian Astrophysical Observatory), Jim Houck (Cornell), and George Rieke (Arizona) made that technology real.

Bob Gehrz (Minnesota), Lawrence "Larry" Manning (Ames), and Charlie Pellerin and Harley Thronson (NASA Headquarters) went above and beyond, being incredibly generous with their time and contacts. Martin Harwit (Cornell) and Dan Weedman (Penn State) provided extremely useful historical perspective.

Although many more people helped make SIRTF possible, I'd like to thank a few who agreed to be interviewed. From Ames: Michael Bicay, Craig McCreight, Ramsey Melugin, and Lou Young. From JPL: Dave Gallagher, Bill Irace, Johnny Kwok, Charles Lawrence, Larry Simmons, and Bob Wilson. From Headquarters: Nancy Boggess, Lawrence "Larry" Caroff, and Frank Martin. From outside of NASA: Eric Becklin (University of California, Los Angeles), Tim Kelly (Ball Aerospace), Marcia Rieke (Arizona), and Tom Soifer (Caltech).

Monographs are not just written by those listed as the author. I am grateful to the staff of the Ames archives, especially Glenn Bugos and April Gage, and the staff of the Headquarters Historical Reference Collection, especially Colin Fries and Jane Odom, all of whom helped me find—and make sense of—the mountains of materials available on a major telescope project.

Many thanks to the people who helped me to reduce that mountain of material into a readable manuscript: Sara Lippincott (Pasadena), editor and fact-checker extraordinaire; Julia Kim (New York University) for archive assistance; and the anonymous peer reviewers who commented on early drafts. At the NASA History Division, thanks to Chief Historian Bill Barry, former Chief Historian Steve Dick, and Steve Garber for shepherding this project through to completion.

To guide my efforts in crafting a history, I thank David DeVorkin (National Air and Space Museum), Patrick McCray (University of California [UC], Santa Barbara), and Robert Smith (University of Alberta) for writing other books that were not only valuable secondary source materials, but excellent models. And I thank my colleague, Kenji Klein (UC Irvine/ California State University, Long Beach), who showed me the possibilities of history as a tool for social science research.

Thanks are also due to a number of talented professionals who helped bring this project from manuscript to finished publication. Editors Emily Dressler and Yvette Smith did excellent work preparing the manuscript for publication. In the Communications Support Service Center (CSSC), Mary Tonkinson, Lisa Jirousek, Chinenye Okparanta, and J. Andrew Cooke carefully copyedited the text; Michele Ostovar did an expert job laying out the beautiful design and creating the e-book version; Kristin Harley thoroughly indexed our work; and printing specialist Tun Hla made sure the traditional hard copies were printed to exacting standards. Supervisors Barbara Bullock and Maxine Aldred helped by overseeing all of this CSSC production work.

Last but certainly not least, I thank my academic colleagues, especially Christine Beckman (UC-Irvine/University of Maryland) for guiding this project as a dissertation and monograph, as well as Barbara Lawrence (UCLA) for introducing me to the SIRTF project team and Cristina Gibson (UC-Irvine/University of Western Australia) for introducing me to Barbara. Thank you so much for all your mentorship.

List of Figures and Tables

Figure 1.1. In Johannes Hevelius's *Firmamentum Sobiescianum sive Uranographia* (Gdansk, 1690), the figure of Orion is drawn as if looking from the heavens toward Earth. Thus, the stars appear as mirror images of ground-based observations. For more details, see Nick Kanas, *Star Maps: History, Artistry, and Cartography*, 2nd ed. (New York: Springer, 2009).

Figure 1.2. The Constellation of Orion (Akira Fujii, Space Telescope Science Institute for NASA, Photo ID STScI-2006-01).

Figure 1.3. "Messier Object 42," illustration of Orion Nebula published in an addition to the first version of the Messier Catalog in the *Mémoires de l'Académie Royale* for 1771 (Paris, 1774), pp. 460–461. The image has been rotated; the original orientation can be viewed online at *http://messier.seds.org/xtra/history/m-m31_42.html* (accessed 30 August 2016); see also *http://messier.seds.org/m/m042.html* and *http://messier.seds.org/*.

Figure 1.4. "M42: The Orion Nebula (widefield)," 2004 (A. Block and R. Steinberg, National Optical Astronomy Observatory, Photo ID NOAO-m42steinberg).

Figure 1.5. "The Infrared Hunter," 2006 (Tom Megeath, University of Toledo for NASA/JPL-Caltech/Spitzer Science Center, Photo ID ssc2006-16c).

Figure 1.6. The Hertzsprung-Russell diagram, reproduced here from "Life Cycles of Stars" (EG-1997(09)-004-GSFC), available at *http://imagine.gsfc.nasa.gov/educators/lifecycles/Imagine2.pdf* (accessed 30 August 2016).

Figure 1.7. Yerkes Observatory staff, May 1946. This photograph is reproduced by permission from the University of Chicago.

Figure 1.8. The spectroscopy experiment and rocket (The Jesse Greenstein Papers, Caltech Archives, Photo ID JLG50.9-1).

Figure 1.9. Jesse L. Greenstein in front of the Palomar Observatory, 1965 (Leigh Wiener, Caltech Archives, Photo ID 10.12-13).

Figure 2.1. Dr. Nancy Grace Roman, NASA's first Chief of Astronomy, with a model of the Orbiting Solar Observatory, c. 1963 (NASA 63-OSO-1).

Figure 2.2. Experiments on the OSO-1 mission, reproduced from J. Lindsay et al., *Orbiting Solar Observatory Satellite OSO I: The Project Summary* (Washington, DC: NASA SP-57, 1965), p. 2, Fig. 1.2.

Figure 2.3. Comparison of wavelengths in the electromagnetic spectrum, reproduced here from *http://science.hq.nasa.gov/kids/imagers/ems/ems_length_final.gif* (accessed 30 August 2016).

Table 2.1. Coolants used in infrared astronomy were compiled by the author. The values for each chemical are from P. J. Linstrom and W. Mallard (2014). NIST Chemistry WebBook, NIST Standard Reference Database Number 69 (National Institute of Standards and Technology, Gaithersburg MD, 2014). Available online at *http://webbook.nist.gov/chemistry/* (accessed 30 August 2016).

Figure 2.4. The telemetry van constructed by Giovanni Fazio and Henry Helmken of the Smithsonian Astrophysical Observatory and used during the 1960s at the National Scientific Balloon Facility in Palestine, TX, to record data from their high-altitude, balloon-borne gamma-ray telescope. Fazio presented this photograph from his personal files at the 37th Committee on Space Research

Scientific Assembly in Montreal, Canada, in a lecture titled "Flying High-Altitude Balloon-Borne Telescopes 50 Years Ago," delivered on Tuesday, 15 July 2008, in session PSB1 (Scientific Ballooning: Recent Developments in Technology and Instrumentation: Reminiscences).

Table 2.2. Reproduced from U.S. Space Science Board, *Space Research: Directions for the Future. Report of a Study by the Space Science Board, Woods Hole, MA*, NRC Pub. 1403 (Washington, DC: National Academy of Sciences, 1966), pp. 346–347.

Figure 3.1. Lead-sulfide (PbS) infrared detectors, c. 1946, constructed by Robert Cashman (National Air and Space Museum, Smithsonian Institution, no. A19940241000).

Figure 3.2. Caltech's 12-foot-high, 62-inch infrared telescope, c. 1965 (National Air and Space Museum, Smithsonian Institution, no. A19820363000). This instrument is now part of the permanent collection of the National Air and Space Museum in Washington, DC.

Figure 3.3. Frank Low's 2-foot-high infrared telescope aboard the NASA Ames Learjet, c. 1972 (National Air and Space Museum, Smithsonian Institution, no. A19830086000). This instrument is now part of the permanent collection of the National Air and Space Museum in Washington, DC.

Figure 4.1. Design for SIRTF as a Shuttle-attached payload, by Fred Witteborn, Lou Young, Larry Caroff, and Eric Becklin at the NASA Ames Research Center, c. 1971, reproduced from "A Short and Personal History of the Spitzer Space Telescope," by Michael Werner, *ASP Conference Series* 357 (2006): 7–22; preprint available at *http://arxiv.org/PS_cache/astro-ph/pdf/0503/0503624v1.pdf* (accessed 30 August 2016). The full caption in Werner's article reads: "Preliminary concept for a far infrared 'liquid helium cooled telescope for sortie mode shuttle,' 1971. The original drawing notes that helium gas vents, a removable vacuum cover, and a 'finder scope with TV' are not shown."

Figure 4.2. Specification of possible SIRTF design and instrument suite, c. 1973, reproduced from the *Final Report of the Space Shuttle Payload Planning Working Groups, Volume 1: Astronomy* (Greenbelt, MD: NASA Goddard Space Flight Center, 1973), p. 8.

Figure 4.3. Design for SIRTF as a Shuttle-attached payload, c. 1980, reproduced from Fred C. Witteborn, Michael W. Werner, and Lou S. Young, "Astrophysics Near-Term Program. Project Concept Summary: Shuttle Infrared Telescope Facility" (Washington, DC: NASA, 1980), Publication NTRS 19830074573, p. 6. This Shuttle configuration is largely the same as one presented on page 670 of Fred C. Witteborn and Lou S. Young, "Spacelab Infrared Telescope Facility (SIRTF)," *Journal of Spacecraft and Rockets* 13, no. 11 (November 1976): 667–674. This article is based on an earlier conference publication, "A Cooled Infrared Telescope for the Space Shuttle: The Spacelab Infrared Telescope Facility," published in January 1976 as AIAA paper #76-174; also see presentation by Lou Young to NASA on the Statement of Work, Phase A for SIRTF. NASA Ames History Office, NASA Ames Research Center, Moffett Field, CA, PP05.04, Larry A. Manning Papers 1967–1988, Box 2, Folder 3.

Figure 4.4. SIRTF is the 1.5-meter telescope on the right in the top drawing and on the left in the bottom one. SIRTF was imagined to be one of several infrared missions that would fly in combination on the Space Shuttle. Images from *Final Report of the Payload Planning Working Groups, vol. 1: Astronomy* (May 1973), p. 29.

Table 4.1. SIRTF as a 1-meter telescope to be flown on the Space Shuttle in 1981. The Space Science Board gave SIRTF the highest priority—as a Shuttle payload. SIRTF is represented at the top of the first column in this table as a liquid-hydrogen (LH$_2$)-cooled 1-meter telescope to be flown on the Shuttle in 1981. Table is from Space Science Board (Richard M. Goody, Chair), *Scientific Uses of the Space Shuttle* (Washington, DC: National Academy of Sciences, 1974), p. 99.

Table 4.2. Members of the Focal-plane Instruments and Requirements Science Team (FIRST), listed in "Appendices to FIRST Interim Report on SIRTF," Ames Research Center, April 14, 1978. This committee defined the initial scientific and technological scope of SIRTF, which formed the basis of NASA's 1983 SIRTF Announcement of Opportunity.

Table 4.3. Timeline of key scientific and technical feasibility reports on SIRTF as a Shuttle-attached payload as of 1980. Source: author.

Figure 5.1. Participants in the first meeting of the Scientific Working Group, held at NASA Ames Research Center, 12–14 September 1984. The names of SWG members are set in bold type. *Front row (left to right):* Giovanni Fazio, George Rieke, Nancy Boggess, Jim Houck, Frank Low, Terry Herter; *back row:* George Newton, Dan Gezari, Ned Wright, Mike Jura, Mike Werner, Fred Witteborn. A copy of this photograph is archived with the NASA History Division's Historical Reference Collection (HRC). NASA Image A84-0569-35.

Figure 5.2. Mobius strip from the *SIRTF Coloring Book* (1982), archived with the NASA HRC.

Figure 5.3. Organizational chart of the NASA Office of Space Science and Applications from the September 1987 SIRTF SWG Meeting Minutes, archived with the NASA HRC.

Figure 5.4. Timeline of the federal budget process as summarized in *http://www.rules.house.gov/Archives/RS20152.pdf*. The most recent version of document RS20152 is available from the Congressional Research Service.

Figure 6.1. Data on projects are compiled from a) the Augustine report, 1990 (available online at *http://history.nasa.gov/augustine/image9.jpg* (accessed 30 August 2016); also b) memo from Charles J. Pellerin, 18 May 1989, on "Process for Center Review/Selection"; and c) *HQ Proposal Ames Review*, 1989, p. 85.

Figure 6.2. Chart based on data from the *NASA Historical Data Book*, vols. I (1959–1968, p. 115), IV (1969–1978, p. 8), VI (1979–1988, p. 522), and VII (1989–1990, p. T-406). Inflation-rate factors are taken from the NASA New Start Index; the most recent version is available at *http://www.nasa.gov/sites/default/files/files/2014_NASA_New_Start_Inflation_Indexuse_in_FY15_final_for_distribution.xlsx* (accessed 9 November 2016).

Figure 6.3. Image is from SIRTF Briefing for OSSA, 22 March 1990, p. 22; copy archived with the NASA HRC.

Table 6.1. From Center Competition Report, 1989; copy archived with the NASA HRC. Scores for Marshall were not ranked by the review committee and are omitted.

Table 6.2. "Full Mission Life Cycle," Figure 7.1 in Dave Doody, *Basics of Space Flight*, JPL D-20120, CL-03-0371; available at *http://solarsystem.nasa.gov/basics/bsf7-1.php* (accessed 30 August 2016).

Figure 6.4. SIRTF Briefing for OSSA, 22 March 1990, p. 22; archived with the NASA HRC.

Table 7.1. "Chronological Changes to SIRTF," compiled by Johnny Kwok, Figure 2 in Robert K. Wilson and Charles P. Scott, "The Road to Launch and Operations of the Spitzer Space Telescope," paper presented at the SpaceOps Conference, Rome, Italy, 16 June 2006; available at *http://trs-new.jpl.nasa.gov/dspace/handle/2014/41102* (accessed 30 August 2016).

Figure 7.1. Graphics compiled by author from SIRTF project documents: the Titan and Atlas models are from "SIRTF Pre-project Review,"

OSSA, 15 May 1992, Jim Evans (JPL); the Delta model is from the 2003 SIRTF launch press kit. A similar compilation appears in Wilson and Scott, "The Road to Launch," 2006.

Figure 8.1. Photo by Russ Underwood (Lockheed Martin Space Systems) of SIRTF during final integration and test at Lockheed Martin, Sunnyvale, CA, available at *http://legacy.spitzer.caltech.edu/Media/gallery/sirtf_04_2002.jpg* (accessed 30 August 2016).

Figure 8.2. Exterior components of SIRTF, line drawing reproduced from M. D. Bicay and M. W. Werner, "SIRTF: Linking the Great Observatories with the Origins Program," in *Origins*, ed. Charles E. Woodward, J. Michael Shull, and Harley A. Thronson, Jr., ASP Conference Series, Vol. 48 (San Francisco: Astronomical Society of the Pacific, 1998), pp. 290–297, available online at *http://adsabs.harvard.edu/full/1998ASPC..148..290B* (accessed 30 August 2016).

Figure 8.3. Cryo-telescope assembly (CTA). Image reproduced from "SIRTF Facility Status Report," presentation by Tom Roellig to the SIRTF SWG Meeting, 16–17 October 1997, Pasadena, CA.

Figure 8.4. Photo of Ball Aerospace technicians working on the telescope's beryllium mirror, available at *http://www.spitzer.caltech.edu/images/2433-SIRTFmirror2-Spitzer-Space-Telescope* (accessed 30 August 2016).

Figure 8.5. Photo of SIRTF instruments installed in the cryostat, Figure 12 in David B. Gallagher, William R. Irace, and Michael W. Werner, "Development of the Space Infrared Telescope Facility (SIRTF)," in *SPIE Proceedings 4850 (IR Space Telescopes and Instruments)*, ed. John C. Mather (Bellingham, WA: SPIE, 2003), pp. 17–29, available at *http://irsa.ipac.caltech.edu/data/SPITZER/docs/files/spitzer/spie_4850_17.pdf* (accessed 30 August 2016).

Figure 8.6. a) Photo of the Infrared Array Camera (IRAC) cryogenic assembly (CA) at Goddard Space Flight Center, available at *https://www.cfa.harvard.edu/irac/00-480c.jpg*. For more detail, see Plate 1 in G. Fazio et al., "The Infrared Array Camera (IRAC) for the Spitzer Space Telescope," *The Astrophysical Journal Supplement Series* 154, no. 1 (2004): 10–17, available at *http://iopscience.iop.org/0067-0049/154/1/10/pdf/0067-0049_154_1_10.pdf* (accessed 30 August 2016). **b)** Photo of the Infrared Spectrograph (IRS) after integration with the Multiple Instrument Chamber, available at *http://irsa.ipac.caltech.edu/data/SPITZER/docs/files/spitzer/spie_4850_122.pdf* (accessed 30 August 2016). For more detail, see Figure 1a in Amanda K. Mainzer et al., "Pre-Launch Performance Testing of the Pointing Calibration and Reference Sensor for SIRTF," in *SPIE Proceedings 4850 (IR Space Telescopes and Instruments)*, ed. John C. Mather (Bellingham, WA: SPIE, 2003), pp. 122–129. **c)** Photo of the Multiband Imaging Photometer for SIRTF (MIPS). Photo credit: Ball Aerospace (by permission).

Figure 8.7. Functional (not reporting) paths are shown. Reproduced from *Spitzer Space Telescope Handbook*, 2010, Figure 2.1; Version 2.1 of this document, published in March 2013, is available at *http://irsa.ipac.caltech.edu/data/SPITZER/docs/spitzermission/missionoverview/spitzertelescopehandbook/* (accessed 30 August 2016).

Figure 8.8. The Spitzer Space Telescope cryogenic telescope assembly (CTA) being prepared for vibration testing, reproduced here from *http://www.spitzer.caltech.edu/images/2430-SIRTFtank-Spitzer-Cryogenic-Telescope-Assembly* (accessed 30 August 2016).

Figure 8.9. Photo of launch at Cape Canaveral. SIRTF is packed inside the white portion of the Delta rocket's nose cone. Photo credit: NASA Kennedy Space Center.

Introduction

Monday, 25 August 2003, Cape Canaveral just past 1:30 in the morning: It's a typical Florida late-summer night—crushingly hot, with humidity so high it might as well be raining. But the starry sky is mostly clear. One cloud squats nearby, threatening to stop the countdown. Yet the small crowd of astrophysicists and engineers isn't looking up. Their gaze is fixed on the launch site a few miles away, where, in a few moments, a rocket will ignite and either carry their telescope into space or blast it into a million pieces.

Some of them have waited almost three decades for this moment. As they labored to build the telescope, their children grew up and started families, their students graduated and became tenured professors. Now the Space Infrared Telescope Facility (SIRTF), their collective progeny, is about to leave the nest, too. Perched atop a 130-foot Delta II rocket with over 500,000 pounds of fuel in its belly, SIRTF waits—and so do its builders.

When the rocket ignites, there is no sound, only a massive fireball that lights up the sky and is reflected in the pond that separates observers from the launch site. A ripple moves across the water's surface as a wave of heat from the launch rushes towards them. Silently, rising out of the smoke, SIRTF moves away from Earth. Just then the roar of the rocket's engines reaches their ears and the shockwave rumbles through their chests, mingling with shouts of joy and relief.[1]

For some members of the astronomy community, this was the start of the story they were looking forward to: the operational phase. While the launch was a climactic moment for those who had worked so long and hard to make SIRTF a reality, it meant that operation of the telescope would now be turned over to other engineers and to the scientists who would share viewing time and access to the data that SIRTF's instruments would collect. The following December, this transition would also be marked by a name change: from SIRTF to the Spitzer Space Telescope, acknowledging the vision and tireless advocacy of the late astrophysicist Lyman Spitzer, Jr., who, in 1946—noting that "such a scientific tool, if practically feasible, could revolutionize astronomical techniques and open up completely new vistas of astronomical research"—first proposed putting a telescope in space.[2]

A technological marvel, the telescope weighed nearly a ton, measured approximately 4.5 meters in length and 2.1 meters in diameter, and carried three state-of-the-art instruments: the Infrared Array Camera (IRAC), the Infrared Spectrograph (IRS), and the Multiband Imaging Photometer for Spitzer (MIPS). It was the most sensitive infrared telescope ever built. SIRTF would relay data that could not possibly be collected from the ground because Earth's atmosphere absorbs most of the infrared radiation emitted by astronomical objects. A space-based infrared telescope would be able to see things no human had ever seen. It

---

1. NASA Kennedy Space Center Press Release 76-03, 25 August 2003, available at *http://www.nasa.gov/centers/kennedy/news/releases/2003/release-20030825.html* (accessed 30 August 2016).

2. John M. Logsdon, ed., with Amy Paige Snyder, Roger D. Launius, Stephen J. Garber, and Regan Anne Newport, *Exploring the Unknown: Selected Documents in the History of the U.S. Civil Space Program*, vol. 5, chap. 3, p. 547, doc. III-1 (Washington, DC: NASA SP-4407, 2001).

would pierce the ubiquitous dust of interstellar space, sending back data on phenomena ranging from planets far beyond the solar system to brown dwarfs too faint to be observed from Earth. It would look back in time to when the universe was only 800 million years old.[3] It bore sophisticated sensors developed by researchers from the military, government, academia, and industry, who had refined the technology for more than 30 years to achieve the kind of sensitivity that would make the invisible visible. And the technology was the easy part.

SIRTF was not yet in development when, in July 1969, a more famous launch took place at the same site. After Apollo 11's successful Moon landing, NASA searched for a new purpose, while an economic recession waited around the corner. In the early 1970s, there was a small group of advocates for an infrared space telescope; however, the field of infrared astronomy was only a few years old, and no one had ever built a space-based observatory of the required complexity. Considering the technical, political, scientific, and economic uncertainties, it was not obvious that a project like SIRTF could—or should—be dared by NASA.

How did SIRTF manage to overcome these uncertainties? The project developed in parallel with the field of infrared astrophysics and was thus unencumbered by legacy or legitimacy.

This monograph makes visible the invisible forces that influenced the design of SIRTF's innovative technology. The lessons learned by the project team over the course of building SIRTF, now better known as the Spitzer Space Telescope, are about managing innovation over time and in the face of uncertainty. These are universal lessons, applicable to any project whose stakeholders control the necessary resources. SIRTF's stakeholders focused on a variety of issues: technical, scientific, political, and economic, as well as organizational needs and goals. What made SIRTF's evolution particularly difficult was that the stakeholders changed over time—in their composition, goals, and influence.

As a machine, Spitzer is an elegant compromise between what was scientifically desirable and what was technically feasible. As a social construction, it was shaped by political, economic, and institutional realities that entailed both possibility and constraint.[4] Moreover, before there could be an operational telescope, stakeholders had to be "sold" on the project—won over by scientists who had not been trained in persuasion, sales, product design, or customer service. They undoubtedly did not know that their telescope would take 30 years to build. It would have been impossible to foresee how issues at NASA

---

3. The Great Observatories Origins Deep Survey (GOODS) provided the first views of the early universe. See M. Giavalisco et al., "The Great Observatories Origins Deep Survey: Initial Results from Optical and Near-Infrared Imaging," *Astrophysical Journal* 600, no. 2 (2004): L93–L98. Data from Hubble, released in December 2003, looked back to a universe 400–800 million years old. Data from Spitzer when the universe was 800 million years old were available by June 2004. For details, see H. Yan et al., "High-Redshift Extremely Red Objects in the *Hubble Space Telescope* Ultra Deep Field Revealed by the GOODS Infrared Array Camera Observations," *Astrophysical Journal* 616 (2004): 63–70.

4. The assertion that social and political forces affect large-scale innovation projects is not new. Interested readers are directed to Wiebe E. Bijker, Thomas P. Hughes, and Trevor J. Pinch, eds., *The Social Construction of Technological Systems: New Directions in the Sociology and History of Technology* (Cambridge, MA: MIT Press, 1987). In the NASA context, see Robert W. Smith, with contributions by Paul A. Hanle, Robert Kargon, and Joseph N. Tatarewicz, *The Space Telescope: A Study of NASA, Science, Technology, and Politics*, with a new afterword (Cambridge: Cambridge University Press, 1993); Brian Woods, "A Political History of NASA's Space Shuttle: The Development Years, 1972–1982," *Sociological Review* 57, no. 1 (2009): 25–46; Stephen J. Garber, "Birds of a Feather? How Politics and Culture Affected the Designs of the U.S. Space Shuttle and the Soviet Buran" (master's thesis, Virginia Polytechnic Institute and State University, 2002).

Headquarters, in Congress, within the economy, and across international politics might affect its design. Eventually, as the project members developed strategies for managing uncertainty and acquiring legitimacy, they came to understand and even anticipate these challenges.

Innovation does not come from individuals; it comes from social interactions among individuals, borrowing ideas and adapting them to a particular setting or project.[5] This is hardly to say that people do not matter. A dedicated core group worked continuously on SIRTF; without them, the project would likely have faltered. There were also individuals who played interim but pivotal roles. Yet the histories of such remarkable human accomplishments often overlook the forces that informed those achievements. The Spitzer Space Telescope was shaped as much by the institutional landscape, the aftereffects of World War II, the Cold War space race, and advances in other scientific fields. Crises also played a role—economic recessions, loss of the Space Shuttle Challenger, and mission failures that cost NASA much of the public's (and the politicians') faith. These forces and events determined, in large part, scientists' access to the ideas and technologies used in their innovations. This monograph traces those invisible threads as they became the warp and weft of Spitzer. It also examines how the team became more adept at that weaving and offers lessons for managers and developers of long-term projects focused on innovative technologies and produced in the face of uncertainty.

In a project of great complexity and duration, there are many aspects one could emphasize, and this study pretends to be neither comprehensive nor definitive. Its purpose is twofold: It serves as a chronology of a major NASA mission that has led to fundamental changes in our scientific understanding of the universe, and it highlights how the development of such a project is impacted by regional and national politics, economic conditions, scientific agendas, and technical requirements. I emphasize the strategies by which the team gained, lost, and regained the resources and legitimacy needed to sustain the project. To help the reader understand those actions, the narrative reaches back in time to when infrared astronomy and NASA were just beginning, with the aim of presenting the Spitzer Space Telescope in its historical context.

In preparing the chronology, I conducted oral history interviews with 29 people, including scientists, engineers, and administrators who participated in the project. I also made use of interviews and publications prepared by historians of science and NASA researchers. I collected meeting minutes of the Scientific Working Group (SWG) that established SIRTF's scientific objectives and preliminary design, as well as the SWG's quarterly reports to NASA Headquarters, feasibility studies, technical reports, and scientific papers relating to the development both of Spitzer and of infrared astronomy. This collection of source materials has been deposited in the Historical Reference Collection (HRC) archives of the History Division at NASA Headquarters in Washington, DC, for use in future scholarly research.

The structure for this monograph is as follows: Chapter 1 outlines the early interest in and obstacles to studying astronomical objects in the

---

5. Academic research refutes the myth of the individual inventor toiling away in solitude; it is the exception, not the rule. All steps of innovation, from conceptualization to diffusion, have been shown to be subject to social processes. Some representative works on this topic include T. Allen, *Managing the Flow of Innovation* (Cambridge, MA: MIT Press, 1997); A. Hargadon and R. Sutton, "Technology Brokering and Innovation in a Product Development Firm," *Administrative Science Quarterly* 42, no. 7 (1997): 16–49; and Pino G. Audia and Christopher I. Rider, "A Garage and an Idea: What More Does an Entrepreneur Need?," *California Management Review* 48, no. 1 (2005): 6–28.

infrared regions. Chapter 2 introduces many of the core participants and institutions that would develop SIRTF. Chapter 3 describes early research efforts along with the challenges and lack of legitimacy that plagued infrared astronomy. Chapter 4 focuses on the technical feasibility studies that provided initial justification for SIRTF, leading to NASA's Announcement of Opportunity, an invitation to the scientific community to develop instruments for the proposed facility. Chapter 5 focuses on how SIRTF, as a NASA project, was managed, including the selection of the Science Working Group and the acquisition of project resources. Chapter 6 focuses on the alignment (and misalignment) of SIRTF with the nation's political goals and those of the scientific community. Chapter 7 addresses the actions taken to reduce costs and attain economic feasibility so as to achieve "New Start" status (i.e., designation as a fully funded project in the NASA budget approved by Congress) in fiscal year (FY) 1998. Chapter 8 details some of the challenges in actually building a one-of-a-kind telescope facility while working across diverse institutions. Chapter 9, the concluding chapter, discusses strategies by which the project was managed despite ever-changing political, economic, technical, and scientific objectives.

CHAPTER 1

Ancient Views

For a world otherwise divided by geography, language, and culture, the stars provide a common point of reference. From almost every place on Earth—even where the Milky Way has been washed out by the city lights of New York, Buenos Aires, Tokyo, Los Angeles, or Paris—one can look up and see the constellation Orion (Fig. 1.1). The stars Rigel and Betelgeuse, which mark Orion's foot and shoulder, are among the brightest in the sky and yet are 640 and 700–900 light-years, respectively, from Earth. Between them lies a trio of stars that compose Orion's Belt. Dangling below the belt are more stars that form Orion's Sword.

Before the advent of infrared astronomy in the 1960s, astronomers thought they knew everything about Orion. This constellation was well known to ancient civilizations: Romans called it the Hunter, and Egyptians the Shepherd. The earliest records of Orion are from Babylon (present-day Iraq), c. 1350–1170 BCE, where the constellation was called Papsukkal, which in Akkadian means "messenger of the gods."[1]

---

1. John H. Rogers, "Origins of the ancient constellations: I. The Mesopotamian Traditions," *Journal of the British Astronomical Association* 108, no. 1 (1998): 9–28; available at *http://adsabs.harvard.edu/full/1998JBAA..108....9R*.

FIGURE 1.1. Orion from Johannes Hevelius, *Firmamentum Sobiescianum sive Uranographia* (Gdansk, 1690).

More than 3,000 years of observation have not dimmed our fascination with Orion. It requires only a bit of imagination to see the Hunter's outline and a dark sky to see his Sword (Fig. 1.2). Away from city lights, the Sword appears to comprise three stars. Upon closer inspection, each of these "stars" turns out to be a cluster of stars.[2] But to the naked eye, the middle "star" in Orion's Sword appears noticeably hazy because it comprises thousands of stars wrapped in a blanket of dust.[3] Thus, when Charles Messier aimed his telescope at Orion's Sword as it hung above Paris on 4 March 1769,[4] he would not have been surprised to find a nebulous cloud filled with multiple points of light. However, he wasn't particularly interested in glowing star formations: Charles Messier was a comet hunter.

In the 18th century, tracking comets was the way to make one's reputation in astronomy. Messier sought to eliminate the spurious accounts of comets that were causing contemporary astronomers much confusion and embarrassment. Too many "comets" were turning out to be nebulae, which, like the hazy star in Orion's Sword, glowed and had fuzzy shapes. Mapping the nebulae was thus a way to separate the stellar background from the comets. Using equipment not much more powerful than modern binoculars, Messier documented more than 100 such objects, including the one now viewable in his telescope, which he duly named the Orion Nebula (Fig. 1.3). He published the first of his careful accounts and illustrations in 1771. It turned out to be so useful to astronomers

FIGURE 1.2. The Constellation of Orion (Space Telescope Science Institute).

that the items he cataloged have become known as Messier Objects. Rather than building a reputation for discovering comets, Messier found lasting fame for classifying the astronomical objects that were a nuisance to him. Today we refer to the Orion Nebula as Messier 42, or simply M42, a designation that derives from its place as the 42nd entry in Messier's *Catalog of Nebulae and Star Clusters*.

---

2. Starting at the tip of the sword and moving upward, these three clusters are NGC 1980 (which includes the iota Ori cluster), M42 (which includes the M43 nebula and the Trapezium cluster), and NGC1977 (which includes two major stars: 42 Ori and 45 Ori). Going a little further up towards the belt, we might also include a fourth "star," which is the more diffuse NGC 1981. For details, see Philip M. Bagnall, *The Star Atlas Companion: What You Need to Know About the Constellations* (New York, NY: Springer, 2012), esp. pp. 330–332; see also Fred Schaaf, *The 50 Best Sights in Astronomy and How to See Them: Observing Eclipses, Bright Comets, Meteor Showers, and Other Celestial Wonders* (Hoboken, NJ: John Wiley & Sons, 2007), p. 150.

3. "NASA's Hubble Reveals Thousands of Orion Nebula Stars," NASA press release (#06–007), 11 January 2006, *http://www.nasa.gov/home/hqnews/2006/jan/HQ_06007_HST_AAS.html* (accessed 30 August 2016).

4. Hartmut Frommert, Christine Kronberg, Guy McArthur, and Mark Elowitz, SEDS Messier Catalog, SEDS, University of Arizona Chapter, Tucson, Arizona, 1994–2016, *http://messier.seds.org/* (accessed 30 August 2016).

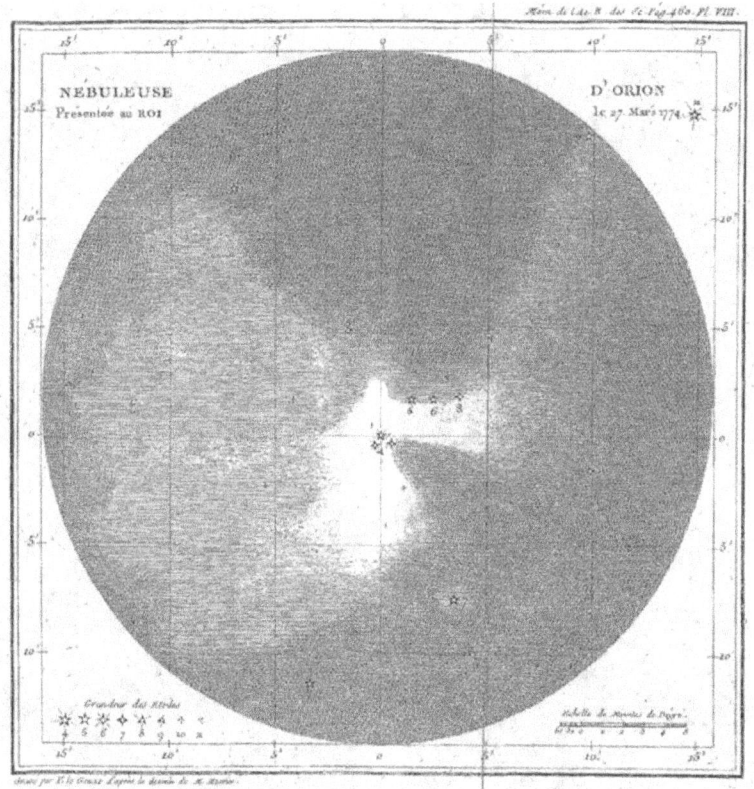

FIGURE 1.3. Charles Messier's drawing of the Orion Nebula M42 (1771), rotated.

More than two centuries later, we are still looking at M42 through telescopes. Images from the ground taken in 2006 reveal its clouds of dust and thousands of stars in great detail (Fig. 1.4, p. 4). Messier's drawing was remarkably accurate and shows the Trapezium, the cluster of stars at the center that is the engine of this energetic nebula. What scientists now understand from optical and infrared observations is that M42 is a stellar nursery (Fig. 1.5, p. 4), where new stars are forming and churning the dust with their stellar winds and gravitation.

Bringing Dust into Focus

Despite the monumental improvements in optical telescopes since Messier's time, dust is still a nuisance for optical astronomers. Starlight, which is simply radiation that reaches Earth's surface, is obscured by dust particles. As light passes through space, it encounters particles from the solid and gaseous remains of ancient stars. The light's path is blocked or bent, depending on the size and shape of the particles. Unfortunately for optical astronomers, dust is everywhere, and it is unevenly distributed.

Very few astronomers considered dust a subject worthy of study before infrared technology came on the scene in the 1960s. Until then, astronomers did as they had always done: They measured the visible wavelengths, in which stars appear as white, blue, yellow, orange, and red, depending on how hot they are. For example, the two brightest stars in Orion are not the same color, even to the naked eye: Rigel is a hot blue supergiant (B class), while Betelgeuse is a comparatively cooler red supergiant (M class). Such color differences help scientists to identify the size and

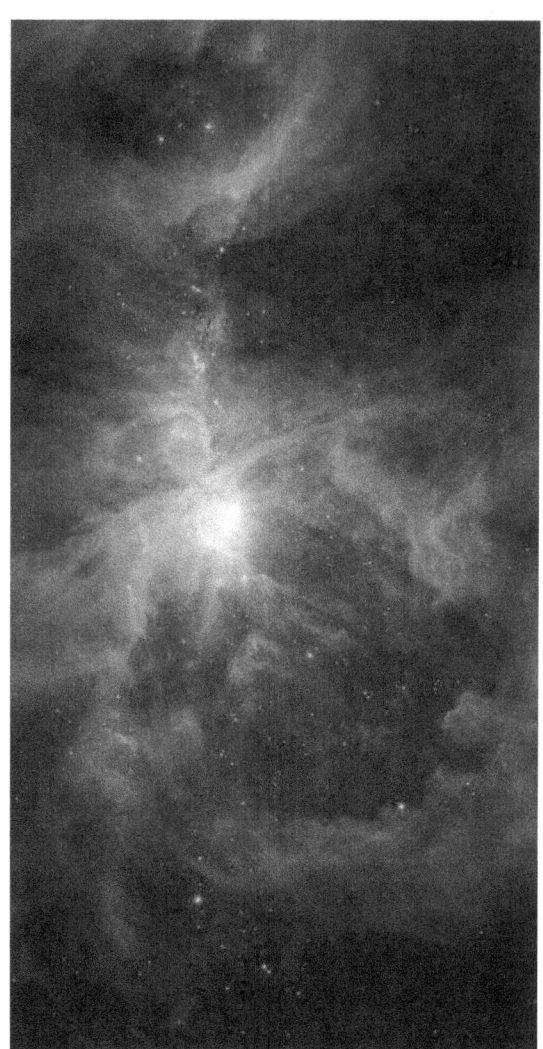

FIGURE 1.4. M42 in the visible spectrum (National Optical Astronomy Observatory).

FIGURE 1.5. M42 in the infrared spectrum (Spitzer Science Center).

temperatures of stellar objects and allow classification of stars as illustrated in the Hertzsprung-Russell diagram (see Fig. 1.6).

Color is a useful way to classify a star, but dust can distort colors. For example, our Sun is a yellow star (G class)—not too hot and not too cold—but when dust is present or sunlight is passing low through the atmosphere at dusk and dawn, the Sun changes from yellow to red. This makes for wonderful photographs but dubious science if results depend on when and where you observe a star.

Dust was an astronomical topic that attracted a select few: "I was interested in dust," Jesse Greenstein said in a 1977 interview.[5] Interstellar dust will redden any light that passes through

---

5. Jesse L. Greenstein, interview by Spencer R. Weart, Session I, Pasadena, CA, 7 April 1977, Niels Bohr Library & Archives, American Institute of Physics, College Park, MD, transcript available at *https://www.aip.org/history-programs/niels-bohr-library/oral-histories/4643-1* (accessed 30 August 2016); see also Jesse L. Greenstein, "Studies of Interstellar Absorption" (doctoral dissertation, Harvard University, Cambridge, MA, 1937).

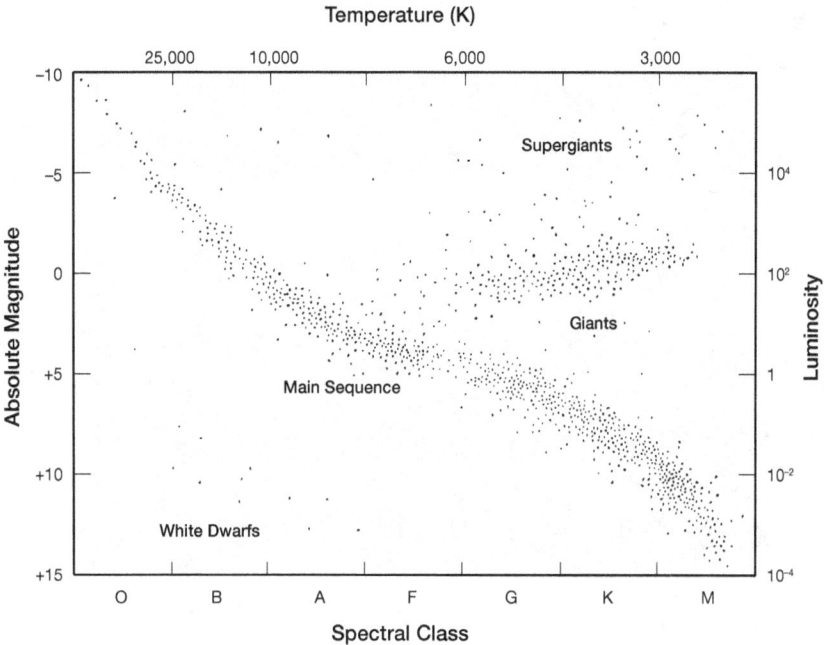

FIGURE 1.6. The Hertzsprung-Russell diagram relates stellar classification to absolute magnitude, luminosity, and surface temperature ("Life Cycles of Stars").

it, and Greenstein's doctoral thesis on interstellar absorption, completed at Harvard in 1937, was one of the first studies to calculate the ratio of light absorption to reddening for the B-class stars. These stars are among the hottest, whitest objects in our galaxy and include Rigel in Orion. Like virtually all other astronomers at that time, Greenstein was an optical astronomer; however, he gravitated toward unusual research topics and was successful despite, or perhaps because of, that predilection. After receiving his doctorate, he was awarded a National Research Council Fellowship that enabled him to continue his work at the Yerkes Observatory.

The View from Yerkes

The Yerkes Observatory was an exciting place to be during Greenstein's tenure there. With its 40-inch refracting telescope, completed in 1895 and still the largest refracting telescope in the world, Yerkes had no trouble attracting top talent (Fig. 1.7, p. 6). Seventy-six miles north of the University of Chicago, which owned it, Yerkes sat in the woods of Wisconsin near Lake Geneva, where the nights were often dark, clear, and cold.

At Yerkes, Greenstein pursued research on comets, but unlike Charles Messier, his interest in comets was incidental to his research on interstellar dust. With access to powerful instruments, Greenstein was able to obtain the first high-resolution spectra of comets: "[W]hen a bright comet came along, here I was with what was then the best spectrograph in the world. There's no reason not to see what the comet spectrum is like on a much bigger scale, just to satisfy curiosity."[6] Greenstein's abundant curiosity led him to undertake research, with Gerard Kuiper, using a highly novel method—rocket-based instrumentation.

---

6. Greenstein interview (Weart, Session III), 26 July 1977.

FIGURE 1.7. Yerkes Observatory staff, May 1946. Seated in the center row are Jesse L. Greenstein (with moustache) and Gerard P. Kuiper (to right of Greenstein), and in the front row is Nancy G. Roman (third from left).[7]

Kuiper had been at Yerkes since 1936, steadily making major discoveries about, among other things, the atmospheres of planets and the moons that orbit them in the outer solar system. In the summer of 1945, as a member of the Alsos mission investigating the level of nuclear energy and nuclear-weapons development in Germany, Kuiper visited a German radar-countermeasures laboratory.[8] He ended up bringing back new ideas for astronomy research after seeing the infrared detectors developed by German engineers for military weaponry. "In fact, [Kuiper] liberated a captured German night-vision infrared device, which he and I experimented with," said Greenstein. "He also liberated what was called a lead-sulfide [PbS] photoconductor cell. With that, he observed the atmospheres of the planets in the near-infrared and was the first to discover that the satellite Titan had an atmosphere. That was done with a formerly classified military device."[9]

---

7. Thanks to Virginia Trimble, who identified some of the people shown here in "Obituary: Jesse Leonard Greenstein (1909–2002)," *Publications of the Astronomical Society of the Pacific* 115, no. 809 (2003): 890–896.

8. Hilmar W. Duerbeck, "German Astronomy in the Third Reich," in *Organizations and Strategies in Astronomy*, vol. 7, ed. André Heck (Dordrecht, Netherlands: Springer, 2006), pp. 383–413; also see Jesse L. Greenstein, interview by Rachel Prud'homme, Pasadena, CA, 25 February–23 March 1982, Oral History Project, California Institute of Technology Archives, transcript available at *http://oralhistories.library.caltech.edu/51/01/OH_Greenstein_J.pdf* (accessed 30 August 2016).

9. Greenstein interview (Prud'homme), 25 February–23 March 1982. According to Dale Cruikshank, although Kuiper was indeed familiar with German technology, the detectors he used were U.S.-made. "The American detectors were declassified in September 1946, and Kuiper soon collaborated with the detector's developer, Robert J. Cashman, on construction of an infrared spectrometer for the study of stellar spectra in the wavelength region of 1–3 micrometers"; see Dale P. Cruikshank, "Gerard Peter Kuiper," in *Biographical Memoirs*, vol. 62 (Washington, DC: National Academies Press, 1993), pp. 258–295, esp. p. 266.

FIGURE 1.8. Spectroscopy experiment and rocket (The Jesse Greenstein Papers).

Kuiper and his colleagues began to apply these German-made infrared detectors to novel types of astronomical research.[10] Similarly, astronomers in New Mexico were finding new uses for the German V-2 rocket. More than 100 German rocket scientists and 300 railway boxcars loaded with V-2 parts were brought to White Sands Proving Ground to continue rocket development under U.S. Army supervision.[11] American scientists were encouraged to supply instruments that could be used for experiments—and ballast—in rocket nose cones during testing. One of the first to answer the call was James Van Allen of Johns Hopkins University's Applied Physics Lab (APL). Established in 1942 to support the war effort, APL produced some of the first pictures of Earth from space in 1946, 12 years before the launch of Sputnik.[12] In a 1982 interview, Greenstein

---

10. Gerard P. Kuiper et al., "An Infrared Stellar Spectrometer," *Astrophysical Journal* 106, no. 2 (1947): 243–250.

11. Michael J. Neufeld, *The Rocket and the Reich: Peenemünde and the Coming of the Ballistic Missile Era* (Cambridge, MA: Harvard University Press, 1996).

12. The earliest photo was taken 65 miles above Earth on 24 October 1946, as reported in Tony Reichhardt, "The First Photo From Space," *Air & Space Magazine*, Smithsonian Institution (1 November 2006), available at *http://www.airspacemag.com/space/the-first-photo-from-space-13721411/* (accessed 30 August 2016). On 7 March 1947, a Naval Research team led by John Mengel mounted a camera to a V-2 that flew 100 miles above Earth. For more details, see White Sands Missile Range chronology at *http://www.wsmr.army.mil/PAO/WSHist/Pages/ChronologyCowboystoV2stotheSpaceShuttletolasers.aspx* (accessed 30 August 2016).

recalled the solar spectroscopy experiment (Fig. 1.8, p. 7) that he had conducted with physicists at APL in April 1947: "I had a spaceflight with a rocket on a captured V-2. It failed. The man who sponsored this project and gave me $7,000 and a V-2 was James Van Allen."[13] Greenstein described this experiment in more detail during a 1977 interview:

> I designed and built a two-quartz prism ultraviolet spectrograph.... We had to build the instrument and the control mechanism. I carried the spectrograph, which was about a meter long, in a wooden box by train to White Sands; saw it mounted; and worked with the Germans, who didn't know how the V-2 behaved at high altitude—I actually met [Wernher] von Braun, with whom I kept up a kind of contact over many years. It was an adventure. But it was a failure. [We didn't understand] just what an experiment in flight does to an instrument.... [U]sing it in a zero-gravity flight environment bound up the spectrography drive. It tried to turn on and tried to turn off, but it never could move. I think it must have stripped the gears. It was really pathetic. But you know, it would have been fun had it worked.[14]

Although the experiment was a failure, it exemplified a new type of science. "I think I lived at a break between when science was an amateur's pursuit and when it was a profession," Greenstein said. "By 1930 or 1934 onward, it became a profession."[15] This new breed of scientists embraced the latest technology. Greenstein added that the work done at Yerkes by Kuiper in the infrared "was perhaps one of the first examples of the increased technological trend in astronomy that began right after the war."[16] The rocket work on the V-2s and other rockets by Van Allen and Greenstein foreshadowed the move toward space and the application of military research to astronomy. As Van Allen observed, "the immense opportunity for finally being able to make scientific observations through and above the atmosphere of the Earth drove us to heroic measures and into a new style of research, very different from the laboratory type in which many of us had been trained."[17]

The ability to use sounding rockets and high-altitude balloons, which had previously been developed by the military for reconnaissance, now made it possible for scientists to study high-altitude physics—solar wind, ultraviolet radiation, cosmic rays—and their effects on weather, communications, and spaceflight. During the 1950s, balloons could reach altitudes of up to 20 miles above Earth, while sounding rockets typically reached 70 miles above Earth, flying 10 times higher than commercial aircraft. It was not long before rockets were developed that could reach 125 miles. In the 1950s, Van Allen developed and flew a balloon-launched rocket, or rockoon: A balloon would hoist the instruments 10 miles up, and then a second-stage rocket would launch them higher, to roughly 60 miles in altitude.

It was also a time of increased government funding for science. A field of new agencies and advisory councils sprouted up. Jesse Greenstein became involved with the National Science Foundation (NSF) at its inception in 1950. James Van Allen was enlisted in 1958 as a founding

---

13. Greenstein interview (Prud'homme), 25 February–23 March 1982.
14. Greenstein interview (Weart, Session II), 21 July 1977.
15. Greenstein interview (Weart, Session I), 7 April 1977.
16. Greenstein interview (Weart, Session II), 21 July 1977.
17. James A. Van Allen, *Origins of Magnetospheric Physics* (Iowa City: University of Iowa Press, 2004), p. 19.

member of the National Academy of Sciences' Space Science Board (SSB), which guided the establishment of NASA.

After the war, many scientists moved around as military projects ended and academic research accelerated. In 1951, Van Allen returned to the University of Iowa, where he had earned his doctorate in 1939, to lead the physics department. Kuiper remained at Yerkes and became its director in 1947, while Greenstein left in 1948 to join the faculty at the California Institute of Technology (Caltech) and build its astronomy department—a plum assignment for someone just 11 years out of graduate school.

As part of his new duties at Caltech, Greenstein was put in charge of the astronomy program, which had prime access to the Hale 200-inch telescope on Mount Palomar (Fig. 1.9). Known simply as the "200-inch," it was the largest telescope ever constructed from the time it began operations in 1949 until 1993, when the 10-meter Keck I, on Mauna Kea, became fully operational.[18] As a scholar at a prestigious university, and with access to the most powerful telescope in the world, Greenstein was highly visible and much sought after. This visibility led to his election to the National Academy of Sciences in 1957 and his participation on many influential committees, panels, and boards. He served as a member of the 1947 Grants Committee for the Office of Naval Research (ONR); chairman (1952) and consultant (1953–1955) to the NSF's Astronomy Advisory Committee; member of the Scientific Advisory Board, U.S. Air Force (1956–1960);

FIGURE 1.9. Greenstein in front of the Palomar Observatory, c. 1965 (Leigh Wiener).

consultant to the RAND Corporation (1957–1961); member of the Astronomy Panel, NASA Steering Committee (1961–1964); and member of the Ramsey Committee (1966), which led to the development of the Hubble Space Telescope.[19] Just as science was now a profession, so the role of the scientist increasingly entailed a large commitment to public service.

Another notable person who departed Yerkes and headed for the public sector after the war was Nancy Roman. After completing her doctorate in 1949 at the University of Chicago, Roman had stayed on as a research associate and then as an assistant professor, conducting research on galactic structure and stellar motion at Yerkes. As a woman pursuing a doctorate in astronomy, she had often encountered discouragement from her parents, college deans, professors, and colleagues. She noted that at the time it was nearly impossible for a woman to get tenure in an astronomy department.[20] So in 1955, she took a job with the Naval

---

18. Although the Soviets built a 6-meter telescope in 1975, design flaws prevented it from exceeding the resolution of the smaller Palomar 200-inch telescope. In 1993, the Keck Observatories in Mauna Kea, Hawai'i, surpassed all previous telescopes with a segmented 10-meter mirror. For an excellent history of Palomar, see Ronald Florence, *The Perfect Machine: Building the Palomar Telescope* (New York: Harper Perennial, 1995).

19. Greenstein interview (Weart, Session II), 21 July 1977.

20. Nancy Grace Roman, interview by Rebecca Wright, Chevy Chase, MD, 15 September 2000, file 3636, NASA History Program Office, NASA, Washington, DC. A 66-page transcript of this interview is available in digital form at *http://www.jsc.nasa.gov/history/oral_histories/NASA_HQ/Herstory/RomanNG/RomanNG_9-15-00.htm* (accessed 30 August 2016); subsequent citations will include page numbers that refer to this transcript.

Research Laboratory (NRL), where work was under way in a new area—radio astronomy—that held great promise and opportunity.

Roman had been at NRL for less than a year when, in 1956, she received an invitation on short notice to attend the dedication of an observatory in Armenia as a guest of the Soviet Academy of Sciences. She had been selected to replace someone who had declined. The director of the new observatory had been impressed by one of her papers, a 1947 two-page research note on so-called high-velocity stars.[21] That paper had been outside of Roman's normal area of interest, and she had all but forgotten it. Surprised by the invitation, yet determined to attend, she recalls:

> I was working for the Navy, still in the middle of the Cold War. I had secret clearance, and I wanted to go to the Soviet Union. It turns out I was the first civilian [working for the military] to go to the Soviet Union after the beginning of the Cold War. So, as you can imagine[,]…this was a bit of a hurdle, and it turned out that the only way I could possibly get the paperwork through was to essentially walk it through myself.… I might add that the paperwork had to go all the way to the secretary of the Navy to get approval for my going, so it was a major undertaking, and…by the time I was finished, people knew me at the Naval Research Lab.…
>
> So I went down to the Soviet Embassy [to get a visa] and sat in the hall for about 45 minutes. Then they took my passport, and they said, "We'll call you when it's ready." [Later] I went down to get it, and this time it was quite a different reception. I was ushered into the office of the scientific attaché, given tea, a cup of tea, and just generally greeted like a great VIP. You know, I was, what, all of thirty-one? It turned out, among other things, that the science attaché had been a translator for the man who was going to be the director of this observatory…and had a lot of respect for him, and the fact that he had invited me to this dedication really put me on a high peg with him.[22]

Roman's tenacity was rewarded. She obtained the visa, as well as approval from the Secretary of the Navy, and, having cleared the bureaucratic hurdles, was present at the dedication ceremony of the Byurakan Observatory on 19 September 1956, where she crossed paths once more with Jesse Greenstein, who had also been invited.[23] Her ability to overcome the naysayers would serve her well as she charted new ground at NASA.

A New Agency

NASA began operations on 1 October 1958, almost one year to the day after the launch of Sputnik by the Soviet Union. The new agency was a combination of other programs, such as the National Advisory Committee for Aeronautics (NACA), the Army Ballistic Missile Agency (ABMA), and parts of NRL. Although Roman was not a member of the NRL group that transferred to NASA, she was well known there, both for her travels and for her technical expertise. "A few months after NASA was formed, I was asked if I knew anyone who would like to set up a program in space astronomy," she remembers. "[T]he challenge of starting with a clean slate to formulate a program that would influence

---

21. Nancy G. Roman, "A Note on Beta-Cephei," *Astrophysical Journal* 106, no. 2 (1947): 311–312.
22. Roman interview (Wright), 15 September 2000, pp. 11–12.
23. Byurakan Astrophysical Observatory Web site, *http://www.bao.am/history.htm* (accessed 30 August 2016).

astronomy for decades to come was too great to resist."[24] Roman expressed her interest in the position, and in February 1959 she joined NASA as head of this new program.

"The first year I was at NASA," Roman said, "I was only responsible for optical and ultraviolet astronomy. Frankly, there wasn't much else."[25] The astronomy effort at NASA was small. Gerhardt Schilling, who was initially in charge, had left NASA by the end of 1959. Roman became the chief of astronomy in the Office of Space Science, a position she would hold under various titles until she retired from NASA in 1979. As the research program grew, it would be carved up into separate offices for x-ray, optical, ultraviolet, and infrared astronomy.[26] But during her first year, Roman managed all of the astronomy-related programs and grants.

Being at NASA posed some challenges. "I had never used the prefix 'Dr.' with my name, but when I started with NASA, I had to," Roman recalled. "Otherwise, I could not get past the secretaries."[27] In the early 1960s, women were rare in scientific fields and rarer still in positions of authority. In many ways, Roman represents the transformations occurring at the time, and she was herself a transformative force in astrophysics research. The way in which astronomy was done had changed radically since the war. During her undergraduate astronomy studies at Swarthmore, Roman recalled meeting with Professor Peter van de Kamp to discuss how she might get started on research: "He was using plates that were taken by his predecessors 50 years earlier, and in turn he felt that he was obligated to replace those with plates that his successors would use 50 years in the future."[28] It was no longer necessary for astronomers to solely rely on what they could find in the archives. There were now alternatives to ground-based telescopes, and observations could be made beyond the visible spectrum. Her exposure to radio astronomy at NRL and rocket-based experimentation at Yerkes had conditioned Roman to welcome new approaches and technology and to be comfortable with change.

---

24. Nancy G. Roman, unattributed interview, NASA Solar System Exploration Web site, *http://solarsystem.nasa.gov/people/romann* (accessed 30 August 2016); see also Roman interview (Wright), 15 September 2000, p. 15.

25. Roman interview (Wright), 15 September 2000, p. 17.

26. Nancy Grace Roman, "Exploring the Universe: Space-Based Astronomy and Astrophysics," in *Exploring the Unknown: Selected Documents in the History of the U.S. Civil Space Program, Volume V: Exploring the Cosmos*, ed. John M. Logsdon, with Amy Paige Snyder, Roger D. Launius, Stephen J. Garber, and Regan Anne Newport (Washington, DC: NASA SP-2001-4407), pp. 501–545, esp. p. 504.

27. Nancy Grace Roman, Women@NASA Web Chat, 4 November 1997, *http://quest.arc.nasa.gov/women/archive/nr.html* (accessed 31 May 2010; .pdf is available in monograph archive in the NASA HRC).

28. Nancy G. Roman, interview by David H. DeVorkin, Washington, DC, 19 August 1980, Niels Bohr Library & Archives, American Institute of Physics, College Park, MD; transcript available at *https://www.aip.org/history-programs/niels-bohr-library/oral-histories/4846* (accessed 30 August 2016).

CHAPTER 2

Getting Infrared Astronomy Off the Ground

At the time of NASA's founding, astronomers were using balloons, sounding rockets, and airplanes to carry their instruments to altitudes between 8.5 and 70 miles above the surface of Earth. All of these airborne methods allow instruments to overcome atmospheric obstructions, such as dust and water vapor, at many (but not all) wavelengths, including parts of the infrared spectrum. The only way to eliminate the interference caused by the atmosphere, especially in the infrared, is to get above it. Instruments on balloon and rocket flights were further constrained by an inability to view a celestial object long enough or accurately enough to make a satisfactory observation (i.e., one that produced interpretable data).[1] Placing an instrument on board a satellite in low-Earth orbit (typically 200–800 miles in altitude) would allow for better observation of sources that often appear faint.

To get into low-Earth orbit, astronomers would need NASA's help—specifically, the help of Nancy Roman, who in 1959 had assumed charge of NASA's observational astronomy program (Fig. 2.1). Although astronomers still observed mostly at optical wavelengths, they were increasingly interested in the high-energy spectra (ultraviolet, x-ray, and gamma ray), which could be viewed only from space. NASA's observational astronomy program included three satellite-based

FIGURE 2.1. Dr. Nancy Grace Roman, NASA's first Chief of Astronomy, with a model of the Orbiting Solar Observatory, c. 1963 (NASA).

---

1. Richard Tousey, "Solar Research from Rockets," *Science* 134, no. 3477 (18 August 1961): 441–448; Martin Schwarzchild and Barbara Schwarzchild, "Balloon Astronomy," *Scientific American* 200, no. 5 (May 1959): 52–59.

observatory programs: the Orbiting Geophysical Observatory (OGO) focused on Earth, the Orbiting Solar Observatory (OSO) focused on the Sun, and the Orbiting Astronomical Observatory (OAO) focused on high-energy celestial phenomena. In a briefing on the OAO held at NASA Headquarters on 1 December 1959, Roman noted that "you cannot exploit the advantages of getting above the atmosphere unless you are able to get up there reasonably large-sized telescopes, and unless you are able to keep these telescopes pointing at one region of the sky for long periods of time to a high degree of accuracy"; she went on to explain that the OAO project would be a prototype for a "stabilized platform system, into which various types of optical instrumentation could be inserted."[2]

Major advances in astronomy made by the 200-inch Palomar telescope alone, since its first light in 1949—such as Walter Baade's 1951 discovery of colliding galaxies and Maarten Schmidt's identification of the first quasi-stellar radio source, or quasar, in 1963—revealed a turbulent and particle-filled universe. Balloons, rockets, and the very first satellites were also showing results, especially in the high-energy spectra.[3] Every discovery only yielded more questions.

In the years following World War II, some startling discoveries were made in the field of astrophysics. From 1952 through 1955, the Office of Naval Research and the Atomic Energy Commission jointly sponsored James Van Allen's investigation of the northern polar aurora's electromagnetic signature by means of rockets.[4] This research resulted in the detection of anomalous radiation surrounding Earth, now known as the Van Allen Belts and understood to be rings of highly charged particles held in place by Earth's magnetic field. The discovery of a blanket of trapped radiation around Earth confounded expectations. It was particularly disconcerting for NASA's piloted spaceflight program, as little was known about the trapped radiation except that it might be lethal to humans and might damage sensitive instruments. Hoping to avoid unwelcome surprises, NASA was interested in sponsoring research on such issues.[5]

Discoveries by experimental high-energy particle physicists also were opening new views. The production of gamma rays in the laboratory led Philip Morrison of Cornell University to predict in 1958 that gamma rays would be found in space.[6] Up to that point, no one had ever looked; gamma rays were first observed in the lab, not in

---

2. Nancy G. Roman, "Transcript of *Orbiting Astronomical Observatories Project Briefing*, NASA, 1 December 1959," NASA TM-X-50191. The full text of this transcript is available online at *http://archive.org/stream/nasa_techdoc_19630040885/19630040885_djvu.txt* (accessed 30 August 2016).

3. For a history of early high-altitude astronomy, see Homer E. Newell, *Beyond the Atmosphere: Early Years of Space Science* (Washington, DC: NASA SP-4211, 1980), esp. ch. 4, n. 7.

4. The first hint of the radiation belts came from data obtained by rockets before 1958, but it was the Explorer 1 mission (1958) that provided definitive evidence of the phenomenon; see Leslie H. Meredith et al., "Direct Detection of Soft Radiation Above 50 Kilometers in the Auroral Zone," *Physical Review* 97, no. 1 (1955): 201–205; James A. Van Allen, "Direct Detection of Auroral Radiation with Rocket Equipment," *Proceedings of the National Academy of Sciences of the United States of America* 43, no. 1 (1957): 57–62; and James A. Van Allen et al., "Observations of High Intensity Radiation by Satellites 1958 Alpha and Gamma," *Jet Propulsion* 28, no. 9 (1958): 588–592.

5. At an OSO project briefing, Dr. James E. Kupperian from NASA's Goddard Space Flight Center noted, "At the moment we are getting new surprises from the Van Allen belt as the data [from other experiments] come in" (*Orbiting Astronomical Observatories Project Briefing*, p. 99). Some of the other experiments that were under way in 1959 are listed in James A. Van Allen, "The Geomagnetically Trapped Corpuscular Radiation," *Journal of Geophysical Research*, vol. 64, no. 11 (1959): 1683–1689.

6. Philip Morrison, "On Gamma-Ray Astronomy," *Il Nuovo Cimento* 7, no. 6 (1958): 858–865.

nature. However, as scientists would soon learn, gamma radiation is present in space but blocked by Earth's atmosphere.

Giovanni Fazio was among the first to look for gamma rays in space. After receiving his Ph.D. in high-energy particle physics at the Massachusetts Institute of Technology, Fazio joined the faculty of the University of Rochester in 1959. There he worked with Professors Everett Hafner, Mort Kaplon, and Joseph Duthie, as well as Joseph Klarmann and Gerald Share, to construct gamma-ray detectors for particle accelerators, such as the Cosmotron at Brookhaven National Laboratory and the Bevatron at the Lawrence Berkeley Laboratory. Fazio came across the 1958 paper by Morrison and a 1959 paper by a Rochester colleague, Malcolm Savedoff, who also asserted that gamma rays could be found in space.[7] "Gamma-ray astronomy was a completely unknown topic at that time," Fazio notes. "Since we were building gamma-ray detectors for the accelerators, we said, 'Wow, we can detect this very easily, so let's get started!'"[8] Fazio would have the same reaction when infrared detectors came along a decade later.

Embarking on a search for cosmic gamma rays, the elementary-particle physics group at the University of Rochester built a Čerenkov counter to look for high-energy phenomena similar to those found in particle accelerators and nuclear reactors. With financial support from the Air Force Office of Scientific Research and NASA, the first instrument was ready for flight in 1959. Testing took place at the Sioux Falls, South Dakota, airport, with the help of Raven Industries, a local start-up that manufactured high-altitude research balloons.[9] However, the experimenters were unable to obtain results due to a balloon failure.[10]

Undeterred, Fazio continued to refine the protocol for balloon experiments, a research platform he would use for 30 years (1959–1989) to study a variety of wavelengths. As for the gamma-ray detector that Fazio and Hafner had designed,[11] another one was built, and Nancy Roman saw that it was placed on board NASA's first observatory in space.[12]

On 7 March 1962, NASA's first Orbiting Solar Observatory (OSO-1) was launched from Cape Canaveral. OSO-1 carried 12 instruments; Figure 2.2 (p. 16) illustrates the placement of several of them in the observatory wheel housing, including Fazio and Hafner's gamma-ray detector.[13] "It was a very crude detector," Fazio notes:

---

7. Malcolm P. Savedoff, "The Crab and Cygnus A as Gamma Ray Sources," *Il Nuovo Cimento* 13, no. 1 (1959): 12–18.

8. Giovanni G. Fazio, interview by author, Cambridge, MA, 26 May 2009. Unless otherwise noted, all interviews were conducted by the author, Renee Rottner.

9. For more on Raven Industries, see *http://ravenind.com/about/our-history/* (accessed 30 August 2016).

10. Giovanni G. Fazio, "Flying High-Altitude Balloon-Borne Telescopes 50 Years Ago," presentation to the 37th Committee on Space Research (COSPAR) Scientific Assembly, held 13–20 July 2008, Montreal, Canada.

11. Giovanni G. Fazio and Everett M. Hafner, "Directional High Energy Gamma-Ray Counter," *Review of Scientific Instruments* 32, no. 6 (1961): 697–702.

12. Technically, the first gamma-ray experiment on board a satellite was launched 3 November 1957, on the U.S.S.R.'s Sputnik 2; however, the results were not released to the international community. In the U.S., the first gamma-ray experiment was Explorer 11, launched on 27 April 1961; for more details, see *http://heasarc.gsfc.nasa.gov/docs/heasarc/missions/explorer11.html* (accessed 30 August 2016). Neither Sputnik 2 nor Explorer 11 was a telescope, let alone an observatory, as they could not be pointed at a source but had to rely on whatever observations could be made while tumbling in orbit. That Explorer 11 did detect 22 gamma rays of cosmic origin was remarkable and is generally considered the beginning of gamma-ray astronomy.

13. Giovanni G. Fazio and Everett M. Hafner, "OSO-1 High-Energy Gamma-Ray Experiment," *Journal of Geophysical Research* 72, no. 9 (1967): 2452–2455; also see Nancy G. Roman, *Orbiting Solar Observatory Satellite OSO I: The Project Summary* (Washington, DC: NASA SP-57, 1965).

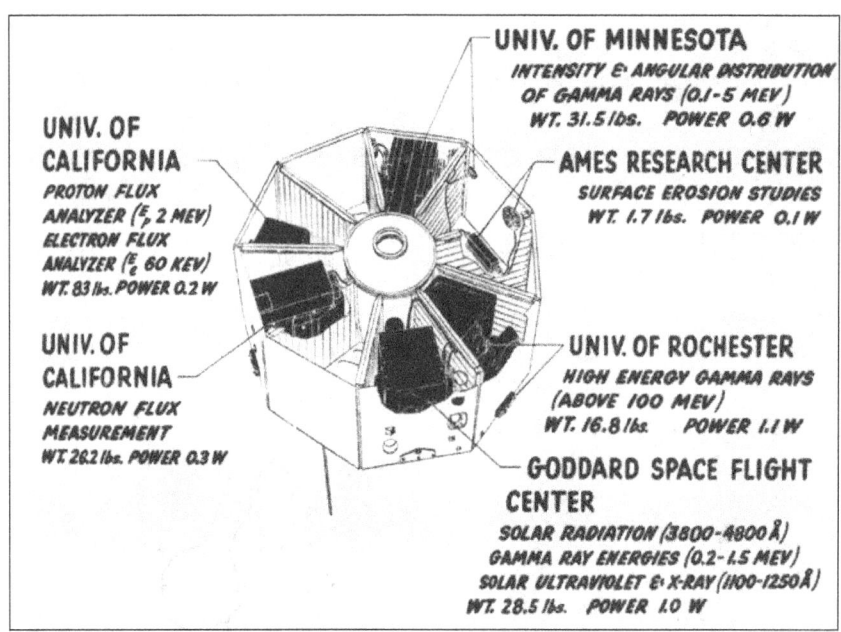

FIGURE 2.2. Experiments on the OSO-1 mission *(Orbiting Solar Observatory Satellite OSO I: The Project Summary).*

I had built it in the laboratory at the University of Rochester—two years from concept to launch. We built it ourselves. The final checkout on the rocket was, I took a voltmeter up to the top of the rocket and checked the voltages—that was the final checkout, you know. It worked, but it didn't detect any gamma rays from the Sun. The camera wasn't sensitive enough. The problem is that there are so few of them. As you go higher and higher in frequency and shorter and shorter in wavelength and more energetic in the spectrum, the flux gets weaker and weaker.[14]

NASA's program of orbiting observatories was very successful, even if the phenomena under study occasionally eluded the investigators. A total of eight OSO missions were launched between 1962 and 1975. The Orbiting Geophysical Observatory program placed six satellites in orbit from 1964 to 1969, and the Orbiting Astronomical Observatory program launched four satellites between 1966 and 1972, of which two succeeded. In total, NASA's three orbiting observatory programs carried aloft more than 200 experiments to examine the cosmos across most of the electromagnetic spectrum, from radio waves to gamma rays. The only wavelength left unstudied was the infrared (Fig. 2.3).[15]

Bringing the Infrared into View

The omission of the infrared spectrum from experiments on the orbiting observatories is glaring. Nearly all of the experiments on the orbiting

---

14. Fazio interview, 26 May 2009.

15. The sole exception was OSO-3, on which Carr B. Neel, Jr., from NASA Ames Research Center, had three instruments designed to measure the long-wave radiation signature of Earth and the testing of materials for use in long-wave sensors. See, for example, Carr B. Neel, Jr., et al., "Studies Related to Satellite Thermal Control: Measurements of Earth-Reflected Sunlight and Stability of Thermal-Control Coatings," *Solar Physics* 6, no. 2 (1963): 235–240.

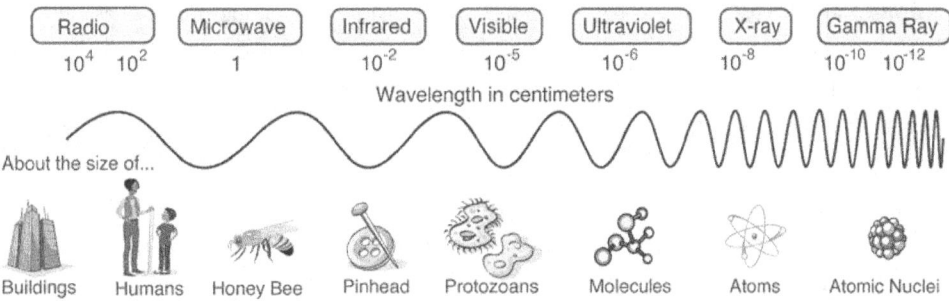

FIGURE 2.3. Comparison of wavelengths in the electromagnetic spectrum (NASA).

observatories were dedicated to the shorter wavelengths: ultraviolet, x-ray, and gamma-ray. At the other end of the spectrum, the longer wavelengths—radio and microwaves—had been a focus of much military research and were used extensively in communications but were not of interest to most astronomers. Scientists originally thought that radio and microwaves were found only on Earth, and only in laboratories. This view changed when, in the 1930s, researchers working on radio communications and equipment observed cosmic radio waves,[16] and in 1965 two physicists working on telecommunications for Bell Telephone Laboratories identified the cosmic microwave background, remnant radiation from the birth of the universe 13.7 billion years ago.[17] Radio and microwave astronomy would follow these discoveries, with the creation of many new, large ground-based telescope facilities, such as the first National Radio Astronomy Observatory (NRAO), at Green Bank, West Virginia, in 1956. Research on infrared, however, lagged behind.

Why was infrared left out of NASA's early space experiments? In 1962 a panel of renowned U.S. astronomers and physicists advised NASA that "infrared observations from satellites of celestial objects should be given a low priority."[18] The committee was convened by the Space Science Board (SSB) of the National Academy of Sciences, the most prestigious scientific society in the United States. The reasoning of the SSB panel was that infrared should continue to be explored by more primitive means, such as balloons and rockets, before being given coveted space on satellites. Ultraviolet and x-ray researchers had proven that they could obtain useful results at

---

16. See Karl G. Jansky, "Radio Waves from Outside the Solar System," *Nature* 132, no. 3323 (1933): 66; and Grote Reber, "Cosmic Static," *Astrophysical Journal* 100, no. 3 (1944): 279–287.

17. See Arno A. Penzias and Robert W. Wilson, "A Measurement of Excess Antenna Temperature at 4080 Mc/s," *Astrophysical Journal* 142, no. 1 (1965): 419–421; and Robert H. Dicke et al., "Cosmic Black-Body Radiation," *Astrophysical Journal* 142, no. 1 (1965): 414–419. At wavelengths longer than the far-infrared, the cosmic microwave background (CMB) glows at 3.5 K and was first detected by Penzias and Wilson in 1965 (for which they received the 1978 Nobel Prize in physics) and first precisely measured in 1992 by NASA's Cosmic Background Explorer (COBE) mission (for which George Smoot and John Mather received the 2006 Nobel Prize in physics). The CMB offers strong evidence of the big-bang theory of the universe, overturning the steady-state models that ruled astrophysics until the 1960s. For more information, two highly readable accounts are Simon Singh's *Big Bang: The Origin of the Universe* (New York: Harper Perennial, 2005) and John C. Mather and John Boslough's *The Very First Light: The True Inside Story of the Scientific Journey Back to the Dawn of the Universe*, rev. ed. (New York: Basic Books, 2008).

18. *A Review of Space Research: The Report of the Summer Study Conducted Under the Auspices of the Space Science Board of the National Academy of Sciences*, National Research Council Publication 1079 (Washington, DC: National Academy of Sciences, National Research Council, 1962), p. 2-25 (hereafter cited as SSB Study Group, *Summer Study*).

higher altitudes and were therefore to be given priority on satellites.

The experts also seemed to think that there was not much to learn from the infrared, particularly in observations of the Sun. The SSB panel reported that "[infrared] observation may not tell us much that is not already deducible from observations in the visible.... [M]uch of the infrared emission from the Sun can be observed from the ground."[19] Such logic assumed that we understood the infrared regime—even though almost nothing was known about the infrared in 1962. The SSB reported to NASA that "there is no broad background of experience on which to base conclusions as to the improvements [in infrared astronomy] to be gained from space observations."[20] In essence, their rather circular argument was that until researchers exhaust all other observational opportunities—however inadequate and obscured by the atmosphere those may be—the panel was against studying the infrared on the grounds that so little was known about it.

The technology to detect infrared radiation was just coming into existence. Lead-sulfide (PbS) detectors, used to sense exhaust heat from enemy jets, had been installed on Sidewinder missiles by 1958. Developed at the Navy's China Lake facility in the Mojave Desert, the Sidewinder was named after an indigenous rattlesnake that hunts by sensing the body heat of its prey. It wasn't until the early 1960s, however, that sensors developed by the military to detect infrared heat—a traceable and therefore important feature of incoming missiles, enemy aircraft, and nighttime troop movements—were not only declassified for civilian use but were also becoming sensitive enough to merit their inclusion on space missions.

A key technical breakthrough came from military and industrial research on semiconductors. Infrared detectors are typically designed around semiconductor crystals, such as silicon, germanium, gallium, indium, or lead, because these elements are differentially sensitive to infrared wavelengths. When a semiconductor is exposed to infrared radiation, it absorbs the photons and becomes capable of conducting a current. It is called a semiconductor precisely because it conducts a current only under some conditions. By tracking these changes in the conductance of the semiconductor, it is possible to measure the presence of infrared light of particular wavelengths.[21] By adding other elements as the semiconductor crystals form (a process known as doping), one can fine-tune the detectors to respond only in the presence of a specific wavelength—like a thermometer that registers when the temperature is at 0°C exactly and not a degree above or below freezing. The research at China Lake and at the Naval Research Laboratory led to new photoconductors made of indium antimonide (InSb), sensitive to the 1- to 5-micron range, and gallium-doped germanium (GaGe), sensitive to 40–120 microns.[22] A mercury-doped germanium (GeHg) detector that even now remains classified was incorporated in 1962 into an infrared instrument by planetary scientist Bruce Murray and engineer James Westphal at Caltech, in collaboration with Dowell Martz at China Lake. As a feasibility study, Murray's team cooled the

---

19. SSB Study Group, *Summer Study*, p. 2-7.

20. SSB Study Group, *Summer Study*, p. 2-24.

21. For a review of Spitzer's infrared technology and how it evolved, see Paul L. Richards and Craig R. McCreight, "Infrared Detectors for Astrophysics," *Physics Today* 58, no. 2 (2005): 41–47; and Frank J. Low et al., "The Beginning of Modern Infrared Astronomy," *Annual Review of Astronomy and Astrophysics* 45 (2007): 43–75.

22. W. J. Moore and H. Shenker, "A High-Detectivity Gallium-Doped Germanium Detector for the 40–120μ Region," *Infrared Physics* 5, no. 3 (September 1965): 99–106; also see Norman Friedman, *The Naval Institute Guide to World Naval Weapons Systems*, 5th ed. (Annapolis, MD: U.S. Naval Institute Press, 2006).

TABLE 2.1. Coolants used in infrared astronomy.

Coolants Used in Infrared Astronomy	Boiling Point		
	Celsius	Fahrenheit	Kelvin
Carbon dioxide ("dry ice")	−79°C	−109°F	195 K
Krypton	−153°C	−244°F	120 K
Oxygen	−183°C	−297°F	90 K
Nitrogen	−196°C	−321°F	77 K
Neon	−246°C	−411°F	27 K
Hydrogen	−253°C	−423°F	20 K
Helium	−269°C	−452°F	4 K
Additional reference points			
Temperature of space	−270°C	−455°F	3 K
Theoretical limit ("absolute zero")	−273°C	−460°F	0 K

Coolants used in infrared astronomy were compiled by the author. The values for each chemical are from P. J. Linstrom and W. Mallard (2014). NIST Chemistry WebBook, NIST Standard Reference Database Number 69 (National Institute of Standards and Technology, Gaithersburg MD, 2014). Available online at *http://webbook.nist.gov/chemistry/* (accessed 30 August 2016).

detector in liquid hydrogen and observed the unilluminated Moon and Betelgeuse in Orion at 8–14 microns, which was most of the infrared spectrum that could be seen from the ground.[23] The results were so surprising that Murray's team was immediately offered the use of the 200-inch Palomar telescope—itself a surprising outcome, for they were not astronomers and viewing time on the 200-inch was in high demand. With some additional modifications to the detector, they obtained the first infrared map of Venus.[24]

Another major technical issue was the emissions from the detectors themselves, a problem known as background-limited infrared performance, or BLIP.[25] The more heat a detector emitted, the more it recorded signals originating from itself rather than from the source of interest. Such unwanted signals, or noise, made it hard to distinguish a faint or faraway celestial source from some part of the detector an inch away. One way to reduce the heat from a detector was to cool the instrument. By placing it in a dewar (a thermos of a supercooled liquid, such as helium, hydrogen, nitrogen, or neon), scientists could ensure that any infrared signals they picked up were not coming from the detector. Observing objects in the far-infrared, such as protostars or debris disks, would require that the detectors be cooled to near absolute zero (see Table 2.1).

By 1968, scientists had identified a range of materials sensitive to infrared across nearly all of its wavelengths. These new materials would allow them to develop detectors that could measure infrared "heat" from relatively cold celestial

---

23. James A. Westphal et al., "An 8–14 Micron Infrared Astronomical Photometer," *Applied Optics* 2, no. 7 (1963): 749–753.

24. Bruce C. Murray et al., "Venus, a Map of Its Brightness Temperature," *Science* 140, no. 3565 (1963a): 391–392. See also the interview with James Westphal, Pasadena, CA, by David DeVorkin, 9 August 1982, Niels Bohr Library & Archives, American Institute of Physics, College Park, MD, *https://www.aip.org/history-programs/niels-bohr-library/oral-histories/24985-1* (accessed 30 August 2016) and Westphal interview, 9 August 1982.

25. For a technical discussion of background limited infrared performance (BLIP), see Stephan D. Price, "Infrared Sky Surveys," *Space Science Review* 142, nos. 1–4 (2009): 233–321.

sources emitting from 1,000 kelvins (K) down to 3 K, just above absolute zero. To appreciate how cold that is, consider that at the freezing point of water (273 K), ice cubes are downright steamy in the infrared.[26]

The rapid pace of semiconductor development during the 1960s led the Space Science Board (SSB) to reverse its 1962 recommendation only three years later. In its 1965 report, the SSB noted that infrared studies should be considered for future orbiting observatories and science programs.[27]

While much had changed in the available technology, the lack of people trained in both infrared techniques and astronomy remained an issue. It was not until 1961 that the astrophysicist Harold L. Johnson, working at the McDonald Observatory of the University of Texas, made the first repeatable, verifiable observations of cosmic sources in the infrared.[28] Johnson is thus, according to some, the first modern infrared astronomer.[29] An early adopter of the new semiconductor materials, Johnson used a lead-sulfide detector to study the near-infrared (typically 1–3 micron[s]), and by 1961 he was using an indium-antimonide detector built by Texas Instruments.[30] During World War II, he had worked with radar-interference techniques; now he applied the knowledge he had gained of photoelectronics to astronomical photometry instruments, earning a doctorate in optical astronomy at the University of California, Berkeley in 1948.[31]

Setting the Stage for SIRTF

The early 1960s was a period of dismal support for infrared astronomy, at least among the astronomers on the influential SSB. Yet it was during this time that others were laying the scientific foundations for SIRTF. Fundamental infrared technology and expertise were aggregating: supercooled low-temperature bolometers at the University of Arizona, infrared spectroscopy at Cornell University, and infrared detectors at the Harvard-Smithsonian Center for Astrophysics.[32] The seeds of these programs were sown in the early 1960s, when the physicists who would develop the technology were converting to infrared astronomy. Two decades later, in 1983, NASA would hold a competition for scientific instruments to be designed for SIRTF. The instruments and Principal Investigators (PIs) for SIRTF would come from these places.

---

26. For reference, 1,000 K = 1,340°F, 3 K = −454°F, and 273 K = 32°F.

27. *Space Research: Directions for the Future. Report of a Study by the Space Science Board, Woods Hole, MA* (Washington, DC: National Academy of Sciences, National Research Council, Publication 1043, 1966), p. 153, hereafter cited as SSB Study Group, *Space Research: Directions for the Future*.

28. Harold L. Johnson, "Infrared Stellar Photometry," *Astrophysical Journal* 135, no. 1 (1962): 69–77.

29. Low et al., "The Beginning of Modern Infrared Astronomy."

30. Johnson, "Infrared Stellar Photometry"; and Harold L. Johnson and Frank J. Low, "Stellar Photometry at 10 μ," *Astrophysical Journal* 139 (1964): 1130–1134. Other work was also under way at Caltech, where, in the summer of 1962, researchers looked at Betelgeuse, Jupiter, Saturn, and the Moon: Bruce C. Murray and Robert L. Wildey, "Stellar and Planetary Observations at 10 Microns," *Astrophysical Journal* 137 (1963): 692–693. At UC Berkeley, R. L. Wax wrote his dissertation on infrared astronomy, publishing it as "Balloon Observations of Infrared and X-ray Intensities in the Auroral Zone," *Journal of Atmospheric and Terrestrial Physics* 28, no. 4 (1965): 397–407.

31. Gérard H. De Vaucouleurs, "Harold Lester Johnson," *Biographical Memoirs*, vol. 67 (Washington, DC: National Academies Press, 1995), pp. 243–261.

32. Although the focus is on the groups that led SIRTF's development, there were a handful of other universities and research labs, such as Caltech and the University of Minnesota, that also had active infrared research programs; for more on this, see Low et al., "The Beginning of Modern Infrared Astronomy."

THE UNIVERSITY OF ARIZONA

In 1960, Gerard Kuiper left Yerkes Observatory and moved to the University of Arizona, in Tucson, where he established the Lunar and Planetary Laboratory (LPL). This was a bold move, as so-called serious astrophysicists at the time focused on stars, not moons and planets.[33] Scientists had shied away from this area since the time of Percival Lowell, who published *Mars* (1895), *Mars and Its Canals* (1906), and *Mars As the Abode of Life* (1908), triggering a slew of sensational stories in the media on Martian civilizations and other science fictions. Embarrassed by the unscientific treatment of this research, few astronomers openly worked on planetary or lunar science. Kuiper's lab was a welcoming refuge for those working on these topics.[34]

NASA was glad to sponsor Kuiper's work. By 1961, the Agency's top priority was the human space program, propelled by President Kennedy's goal to land an American on the Moon by the end of the decade.[35] NASA was eager for any research that could help with this task and provided Kuiper with funds for lunar studies and telescopes.

Kuiper began to build up the new department at Arizona. Over the next four years he hired both Harold Johnson and Frank Low. Kuiper had worked with Johnson at Yerkes (and briefly at the University of Texas) and was impressed by Johnson's work with detectors. In Kuiper's lunar lab, Johnson "was free to pursue his main line of interest—namely, the infrared photometry and interference spectroscopy of stars," a former colleague noted. "He used to joke that he was the 'stellar division' of the Lunar and Planetary Laboratory."[36] Frank Low was a low-temperature-physicist-turned-infrared-astronomer and would provide the essential know-how for cooling infrared detectors to a few degrees above absolute zero. Low recalls his conversion while working with semiconductor materials:

> [I]n my first professional job at Texas Instruments (TI) Central Research Lab in Dallas, I became interested in developing a modern version of a cryogenically cooled bolometer perfectly suited for exploring the spectral range from 1 μm to 1.2 mm. In early 1961 I published an article that described a novel way to measure [infrared] radiation by using basic bolometer principles. My paper explains in full detail how the new germanium device functions [*Low*, 1961]. When the article finally appeared in print I was greatly surprised by its positive reception among astronomers.... After the article appeared, several eager astronomers visited me in Dallas. Among the visitors [was] a young graduate student, Carl Sagan. He was eager to have me build a bolometer system so NASA could fly an [infrared] spectrometer on a balloon to look for organic molecules in a search for life on Mars. The system needed both the detector

---

33. Joseph N. Tatarewicz, *Space Technology and Planetary Astronomy* (Bloomington, IN: Indiana University Press, 1990).

34. Cruikshank, "Gerard Peter Kuiper," p. 273.

35. Kennedy announced the goal of putting a person on the Moon on 25 May 1961. However, two months earlier, the SSB recommended that *science* be the prime motivation for NASA's activities; they did not think that the focus should be on human spaceflight; see letter from Lloyd Berkner, chairman of SSB, to James Webb, NASA Administrator, "Policy Positions on (1) Man's Role in the National Space Program and (2) Support of Basic Research for Space Science," 27 March 1961, Space Science Board, National Academy of Sciences, National Research Council, http://www.nap.edu/catalog/12427/policy-positions-on-1-mans-role-in-the-national-space-program-and-2-support-of-basic-research-for-space-science-march-27 (accessed 30 August 2016).

36. De Vaucouleurs, "Harold Lester Johnson," p. 251.

and a liquid helium dewar to hold it just a few degrees above absolute zero.[37]

With nascent technologies and low-temperature techniques used by only a few infrared physicists, many established astronomers considered infrared a dead end, while the physicists had no training in astronomy. Those who did pursue infrared astronomy often had to cross professional boundaries and endure skepticism from their colleagues. In a history of infrared astronomy, astrophysicist Martin Harwit writes:

> The difficulties of making observations with liquid-helium-cooled devices at that time are hard to grasp today. Observatories were not equipped with liquid helium, helium transfer lines, vacuum pumps for pumping the helium down to lower temperatures, or most of the necessary electronic instruments. An observer had to arrive bringing all this equipment along. The night assistants in charge of the telescopes, who had never seen anything like this and didn't like what they saw, had to be mollified, if not by the observer then by the site director.[38]

In July 1963, before Low came to the Lunar and Planetary Laboratory, he had worked with Johnson on an infrared project at McDonald Observatory, aided by Arnold Davidson, a radio astronomer from the National Radio Astronomy Observatory in Green Bank, West Virginia. The skepticism of the astronomy establishment regarding infrared astronomy was on display during that experiment. Davidson and Low were using an instrument they had built from indium-antimonide detectors developed by the U.S. Department of Defense (DOD) and cooled to super-low temperatures in the technique recently developed by Low. Johnson had arranged to test their equipment at the observatory. Low recalls an encounter there as they waited for the skies to clear:

> [T]he dome at McDonald Observatory could be reached only by foot on rather narrow and steep steps. One day, as Arnold and I were leaving the dome, we crossed paths with a distinguished looking optical astronomer who was surprised to find two young persons there. He wanted to know who we were and why we were there. We identified ourselves and told him we were working with Harold Johnson in the infrared. His response was that he could not understand why we were spending our time waiting for clear weather when we were not going to get any results in the infrared.[39]

Johnson, Davidson, and Low succeeded beyond even their own expectations. The results they achieved were foundational to the nascent field of infrared astronomy. Davidson and Low had covered the spectral range from 8 to 14 microns, and their data supported Johnson's typology of infrared-source temperatures (a sort of Hertzsprung-Russell diagram using heat rather

---

37. Low et al., "The Beginning of Modern Infrared Astronomy," pp. 44–45. The paper Low refers to in this passage is Frank J. Low, "Low-Temperature Germanium Bolometer," *Journal of the Optical Society of America* 51, no. 11 (1961): 1300–1304. At the time of this visit to Low, Sagan was not a student but a postdoc on a Miller Fellowship (1960–1962) at Berkeley. Sagan had received his doctorate from the University of Chicago in 1960 under the supervision of Gerard Kuiper. The detector was for the Stratoscope II spectrometer project at Berkeley, details of which can be found in R. E. Danielson et al., "Mars Observations from Stratoscope II," *Astronomical Journal* 69, no. 5 (1964): 344–352.

38. Martin Harwit, "The Early Days of Infrared Space Astronomy," in *The Century of Space Science*, vol. 1, ed. Johan A. M. Bleeker, Johannes Geiss, and Martin C. E. Huber (Dordrecht, Netherlands: Kluwer, 2001), pp. 301–330, esp. p. 305.

39. Low et al., "The Beginning of Modern Infrared Astronomy," p. 46.

than light), a framework by which the distances of objects are still measured.⁴⁰

The new cooled instrument also made technological leaps. "The linearity, repeatability, and sensitivity were all much higher than one would expect for an all-new instrument," Low noted.⁴¹ It often takes several hours of continuous observation to get a high-resolution image of a celestial source in the visible wavelengths. In contrast, Davidson and Low measured infrared light from Mars, Jupiter, Saturn, Titan, and 24 stars—in a single night.

CORNELL UNIVERSITY

Two thousand miles away from Tucson, a similar transformation was taking place at Cornell University, where another physicist was converting to infrared astronomy. Martin Harwit had earned a doctoral degree in physics from MIT in 1960 but had done graduate work at the University of Michigan in atmospheric research and postdoctoral research at Cambridge University with the famous astrophysicist Fred Hoyle.⁴² With this exposure to physics, astronomy, and atmospheric research methods, Harwit arrived at Cornell in 1962. He joined the Center for Radio Physics and Space Research, under the leadership of Thomas Gold, with the goal of working in the infrared using telescopes mounted on sounding rockets. This was an audacious goal in the early 1960s, given the state of rocketry and sensors—and exactly the sort of unconventional research that Gold relished.

To supplement his knowledge, Harwit contacted Herbert Friedman at the Naval Research Laboratory and was invited to spend a year there, supported by a National Science Foundation fellowship. Working with Henry Kondracki, Harwit set up a lab to perform rocket-borne infrared astronomy. He recalls that after returning to Cornell from his tutorial year at NRL:

> I got a phone call from Nancy Roman, who was the head of Astrophysics at NASA at the time. She asked whether I would want to set up a program of infrared astronomy at Cornell—which was sort of astonishing, in a way. I told the head of my department, Tommy Gold, and he spent half an hour trying to figure out why they should have asked me to do that when he was having trouble getting money from NASA. But he was continually criticizing NASA, and they were a little bit vindictive in those days.... I got a dowry of $250,000 to set up a lab. [Roman] asked me what would it cost, and because we had set up this lab at NRL I knew what I needed. So I said $250,000 and $100,000 a year after that. So that's what they gave us and that allowed us to fly rockets maybe every six months if we could recover a rocket payload, and once a year if we had to start from scratch—which unfortunately happened a lot because rockets in those days were very unreliable.... It was a really heartbreaking operation.... [T]he rocket would be spinning the whole time and you couldn't get anything [because the equipment couldn't deploy], or the parachute would break off. It was just one thing after another.⁴³

---

40. Low et al., "The Beginning of Modern Infrared Astronomy," passim; for the published typology, see Harold L. Johnson, "Astronomical Measurements in the Infrared," *Annual Review of Astronomy and Astrophysics* 4 (1966): 193–206.

41. Low et al., "The Beginning of Modern Infrared Astronomy," p. 47.

42. Martin O. Harwit, interview by David DeVorkin, Washington, DC, 20 June 1983, session I, Niels Bohr Library & Archives, American Institute of Physics (AIP), College Park, MD, *https://www.aip.org/history-programs/niels-bohr-library/oral-histories/28169-1* (accessed 30 August 2016).

43. Martin O. Harwit, interview by author, Cambridge, MA, 26 May 2009.

The failure rate of rocket-borne experiments was high, whether due to the rockets themselves or the cold temperatures in the upper atmosphere, which often froze the instruments before any data could be collected. Given these hazards, it was risky to trade the relative comforts of a cold and isolated ground-based observatory for the erratic problems of airborne instruments.

Although NASA was supportive of high-atmosphere research using rockets, the astronomy community in the 1960s was slow to embrace the new technology. For nearly four centuries, the paradigm for astronomy was to use a ground-based telescope, with the largest mirror one could construct, set inside an observatory on the highest possible hilltop. The observatory would serve multiple experimenters, ideally providing them with operational and data support. Observations could be sustained or repeated over long periods. In contrast, rocket-mounted experiments captured only a few minutes of data—and only a single cross-sectional slice, as the vehicle ascended through the upper atmosphere. When the rocket reached apogee, it slowed and fell back to Earth, arriving (it was hoped) in one piece with the data on board or already telemetered to receivers on Earth. Those "receivers" were not the same sophisticated and automated tracking systems employed today; often they were simply

FIGURE 2.4. The challenges of capturing data: gamma-ray telemetry in the 1960s (Giovanni Fazio, personal files).

graduate students chasing the signal using homemade equipment, as shown in Figure 2.4.

In the infrared, these limits on the data were offset by the rare opportunity to get above the atmosphere and view parts of the infrared spectrum unobservable from Earth. Even one minute of data collected on a rocket was more than could be collected from the ground by any observatory. Thus, while the risks of failure were high, so were the scientific rewards.

The costs associated with rocket-borne experiments were modest compared with the rest of NASA's budget for space science, which averaged $500 million annually throughout the 1960s (about 11.5 percent of NASA's total annual budget).[44] The total funds available, across all of the government agencies that supported rocket-based science, amounted to $50 million, half

---

44. Advisory Committee on the Future of the U.S. Space Program (Norman Augustine, chair), *Report of the Advisory Committee on the Future of the U.S. Space Program* (Washington, DC: NASA, 1990), hereafter cited as "Augustine report," Figure 7.

TABLE 2.2. Typical costs of rocket-borne experiments (in 1966 U.S. dollars).

Rocket	Costs (in 1966 dollars)	Pointing System
Aerobee meteorological rocket	$10,000 per experiment (e.g., photographing vapor trails in atmosphere)	None
Aerobee meteorological rocket	$20,000–40,000 per experiment (e.g., ionospheric measures)	None
Aerobee meteorological rocket	$150,000 per experiment (e.g., galactic x-ray mapping)	None
Aerobee-Hi	$200,000 per experiment	Biaxial
Aerobee-350	$500,000 per experiment	Multidirectional

of which was provided by NASA. A typical experiment would cost $50,000 to $200,000 (in 1966 U.S. dollars), including launch, equipment recovery, and data reduction, as shown in Table 2.2.

Harwit had the funds to conduct rocket-borne astronomy; now he needed a team to help solve the many technical problems. He hired Henry Kondracki, the mechanical engineer with whom he had worked during his year at NRL.[45] In 1967, James R. Houck joined the Cornell group, first as a postdoctoral fellow and later as a professor. Houck had been a doctoral student in solid-state physics at Cornell, working in the building just next door to Harwit. As yet another physicist who converted to infrared astronomy, Houck would become the PI for SIRTF's infrared spectrograph in 1984. One of Harwit's graduate students, Michael Werner (Ph.D., 1968), would become SIRTF's project scientist and, by most accounts, the force that held it together for three decades. Judith Pipher (Ph.D., 1971) and B. Thomas Soifer (Ph.D., 1972) also began working with the infrared group at Cornell and would each play a role in SIRTF.

Meanwhile, Harwit pursued the infrared in other ways. It was only a small shift to consider putting the equipment on aircraft, especially after NASA cut off the funding for rockets in the early 1970s due to decreases in Agency funding.

NASA was willing to fund this new work and was already funding Frank Low, who had pioneered airborne infrared astronomy by installing an infrared telescope on a Learjet. Although the cooled instruments could not fly as high, they could take measurements over a longer period of time, and the Learjet, unlike a rocket, was a reusable platform, making it cost-effective for gathering data. One of the first observational targets was an astronomy favorite—the Orion Nebula. Harwit recalls the early work:

> Houck developed a compact, fully liquid-helium-cooled, grating spectrometer for the mid-infrared range that was sufficiently small to be mounted on the 30-cm (12-inch) telescope on the NASA Learjet. With this he obtained spectra of Jupiter and the Orion Nebula...across the 16–40 μm range.... Using a slightly modified copy of this design on the Learjet, Dennis B. Ward and [Harwit] obtained a first spectrum of the Orion Nebula from ~75 to 100 μm....

Infrared astronomy was coming into its own as an analytical tool but there was still a great deal that was totally unknown: No unbiased survey of the sky had been made at long infrared wavelengths, so that observers

---

45. Harwit, "The Early Days of Infrared Space Astronomy," p. 309.

continued to primarily work with sources familiar from visible or radio observations. We had no idea of what else might be found if the methods were at hand. The infrared background radiation also remained a complete mystery.[46]

HARVARD-SMITHSONIAN CENTER FOR ASTROPHYSICS

Giovanni Fazio, the elementary-particle physicist who became a gamma-ray astronomer, would also soon convert to infrared experiments. In 1962, after putting a gamma-ray detector on NASA's OSO-1 satellite, he left the University of Rochester and joined the Smithsonian Astronomical Observatory (SAO) in Cambridge, Massachusetts.[47]

To help with his balloon-based astronomy program at SAO, Fazio hired George Rieke, a graduate student of the Harvard experimental physicist Jabez Curry Street. There had been a long historical association between Harvard's Department of Astronomy and the Smithsonian Astrophysical Observatory, where many Harvard faculty had joint appointments. Thus, in 1973, the two institutions formed a single organization and named it the Harvard-Smithsonian Center for Astrophysics (CfA), with which Fazio became affiliated.

Rieke worked with Fazio to improve the gamma-ray detectors. Gamma rays, as noted, are rather rare events, and no matter how good the equipment, it is often difficult to detect them in sufficient quantities on which to draw scientifically valid conclusions. Fortunately for Rieke, the experiments were successful enough to form the basis of his 1969 dissertation on detecting gamma rays in space.[48] After graduating from Harvard, he was hired by Gerard Kuiper and joined the astronomy faculty of the University of Arizona, where he would eventually be selected as PI for SIRTF's Multiband Imaging Photometer (MIPS).[49] Fazio would be selected as PI for the Infrared Array Camera (IRAC).

By 1970, Fazio had become dissatisfied with using balloons to search for gamma rays and turned instead to search for infrared sources. In a 2009 interview, he recalled that "the number of photons being detected divided by the number of hours that I was spending working on it, was so small that I quit... I had had it."[50] Although the search for gamma rays from balloons and rockets was declared a bust,[51] Fazio considers the time well invested for what was to come:

> I find that people say, "You wasted all that time." I really didn't waste time. I mean, I learned a lot of things. I learned ballooning. Without the balloon, I wouldn't have gotten into infrared. Everything I've ever done, even though it's remote, in some ways helped in the future. I've never found anything that didn't help me in the future.[52]

---

46. Harwit, "The Early Days of Infrared Space Astronomy," pp. 315–316. "Infrared background radiation" is a reference to the cosmic microwave background.

47. Fazio interview, 26 May 2009.

48. George H. Rieke, "A Search for Cosmic Sources of 10 Exp 11 TO 10 Exp 14 EV Gamma-Rays" (doctoral dissertation, Harvard University, 1969).

49. George H. Rieke, interview by author, Pasadena, CA, 9 June 2009.

50. Fazio interview, 26 May 2009.

51. R. K. Sood, "Detection of High Energy Gamma-rays from the Galactic Disk at Balloon Altitudes," *Nature* 222, no. 5194 (17 May 1969): 650–652.

52. Fazio interview, 26 May 2009.

To help make the switch from gamma rays to infrared, Fazio was joined by his doctoral student, Edward L. (Ned) Wright (Ph.D., 1976) and by Tom Soifer, who did postdoctoral research in 1973 after finishing his doctorate at Cornell. Also doing graduate work at the CfA was Michael A. Jura (Ph.D., 1971). Wright and Jura, as faculty at UCLA, would later join SIRTF as interdisciplinary scientists. Soifer would go on to direct the Spitzer Science Center, which is where the science operations team processes the raw data from the observatory, and manages requests for viewing time and access to archived datasets.

Ready for an Infrared Revolution

Even when infrared experiments were successful, the evidence they provided remained doubtful. Because the methods were so unconventional and prone to failure, it was hard to differentiate significant results from spurious ones. In many cases, results could not be verified by ground-based telescopes, nor did the results fit prevailing theory. One way to interpret this is that the airborne infrared astronomers had found entirely new phenomena—so, naturally, the data from the new instruments and theory did not match. A simpler interpretation is that the data from the balloon and rocket experiments were wrong.

In 1967, three separate papers appeared in the astronomical journals that made traditional astronomers take notice of the new infrared astronomy. Kris Davidson and Martin Harwit provided a theoretical paper that suggested infrared emissions from young massive stars, born inside dusty clouds and still in their formative stages, should be observable at infrared wavelengths long before their cocoon of dust clears and the stars become visible at optical wavelengths.[53] Using the biggest ground-based telescope—the 200-inch Palomar—and fitting it with an infrared detector, Caltech's Gerry Neugebauer and his graduate student Eric Becklin took observations of the near-infrared wavelengths from 1.5 to 13.5 microns.[54] It was a technical accomplishment, and also a scientific one. Scientists were astonished at their finding of a very strong infrared source within the Orion Nebula—Orion still had secrets despite centuries of astronomical observations. The source, thought to be a protostar buried in a cloud of gas and dust, was not visible in the optical wavelengths and thus had not been detected before. To verify this finding, Frank Low and his graduate student Doug Kleinmann at the University of Arizona, using a different telescope and detector, took a look at a longer infrared wavelength of 20 microns. Not only did Low and Kleinmann confirm the Becklin-Neugebauer object, they also found another previously undetected source in the Orion Nebula, even cooler than the first and subsequently named the Kleinmann-Low nebula.[55]

With verification through repeated experiments and theory to support the findings, traditional astronomers began to see that the infrared might be a valuable wavelength to study all on its own. Together with the technology suitable for studying infrared and people with the expertise to adapt it to astronomy, the pieces for building SIRTF were falling into place. George Rieke would later observe that

> SIRTF was built when infrared astronomy was still in the pioneering stage. The people who had to figure out how to build

---

53. Kris Davidson and Martin Harwit, "Infrared and Radio Appearance of Cocoon Stars," *Astrophysical Journal* 148 (1967): 443–448.

54. Eric E. Becklin and Gerry Neugebauer, "Observations of an Infrared Star in the Orion Nebula," *Astrophysical Journal* 147 (1967): 799–802. This was a follow-up study to Neugebauer's infrared sky survey at 2 μm using the Mt. Wilson 60-inch telescope: Gerry Neugebauer et al., "Observations of Extremely Cool Stars," *Astrophysical Journal* 142 (July 1965): 399–401.

55. Douglas E. Kleinmann and Frank J. Low, "Discovery of an Infrared Nebula in Orion," *Astrophysical Journal* 149 (July 1967): Letters, L1–L4.

the first decent infrared astronomy instruments were the ones who were involved in building SIRTF. Usually it's a much longer time between when a field is invented and when you get to build a really ambitious space project. And so you end up getting specialists in the science and specialists in the technology. The technology specialists eventually become engineers instead of scientists. So there is a sort of middle-aging of fields before they get into the space game, in general, and that didn't happen with the infrared—we got in early.[56]

---

56. Rieke interview, 9 June 2009.

CHAPTER 3

Making the Case for SIRTF

The year 1967 marked a turning point in the study of infrared astronomy. Astronomers had been pointing telescopes at the Orion Nebula for hundreds of years and were now stunned to find two entirely new nebulae embedded there, which became known as the Kleinmann-Low and Becklin-Neugebauer objects. Looking through the infrared window had opened new horizons, making it clear that an infrared universe existed and that our knowledge of it was rudimentary at best. Earlier rocket-borne experiments had indicated that there were many new things to observe in the infrared, but such results were so unexpected that astronomers mostly dismissed them as instrument error.[1] Skepticism was compounded by the fact that much of the research was not done within the astronomy community, but by outsiders: physicists in concert with the military. It was not until ground-based observations confirmed these phenomena—using traditional telescopes modified with infrared detectors—and physicists replicated the rocket-based results at several university observatories that infrared astronomy began to be taken seriously. These data forced traditional astronomers to reevaluate what they knew about galaxies, stars, and planets and to speculate about unknown objects that might be observable only in the infrared.

Telescope Surveys and Telescope Facilities

The first step in understanding the infrared universe was to conduct a sky survey. Sky surveys are comparable to the mapping expeditions of Lewis and Clark, where the goal was to determine the lay of the land. Once a rough sketch was made, other teams could return to make more detailed observations. For the sky in the optical wavelengths, many surveys already existed, including Messier's catalog of 1771. The infrared was terra incognita.

For astronomers, it made little sense to build an elaborate infrared telescope that could be pointed at celestial targets until they understood what sources existed in the infrared wavelengths and were worth observing. A great deal could nevertheless be accomplished with a fairly simple telescope. First, all that was needed was a map of the infrared region. To construct it, infrared instruments were being strapped to ground-based telescopes, balloons, rockets, aircraft, and—the

---

1. Harwit interview, 26 May 2009; and James R. Houck, interview by author, Ithaca, NY, 25 May 2009.

newest launch vehicle under development—the Space Shuttle.

Giovanni Fazio, the Harvard physicist who had previously built a gamma-ray detector for NASA's OSO satellite, now was designing an infrared telescope (IRT) to fly on the Shuttle, which was still under development; the IRT would serve as a test bed for space-based infrared sensor technology. The IRT would also provide useful data on the operating conditions aboard the Shuttle and what effect these might have on infrared observations.[2] But the IRT would not be launched until the 1980s, so it was not a substitute for other, more immediate infrared survey efforts; nor was it a substitute for an instrument like SIRTF. While the letters IRT stand for *infrared telescope* in both cases, the last letter in SIRTF stands for *facility*. As a facility, SIRTF would be more than a telescope; it would have features to which astronomers were accustomed, such as a pointable telescope and multiple filters and instruments for running different experiments. Modern ground-based observatories often have several telescopes of various sizes and with sensitivity to different wavelengths. Similarly, SIRTF would carry a variety of instruments, giving astronomers flexibility in designing experiments. And like its ground-based counterparts, SIRTF would be capable of zooming in with great precision on particular targets and holding them in view for extended periods. By contrast, a sky survey could (and arguably should) be conducted using a comparatively simple, single-purpose telescope.

Leveraging Military Research

While astronomers stood to gain the most from a sky survey, few were interested in doing what was essentially engineering work. In contrast, military researchers and university physicists were willing to develop the technology for an infrared sky survey. The Air Force had determined by the late 1950s that an infrared sky survey was needed, but not for the purpose of astronomical research; tacticians wanted to be able to distinguish the heat of stellar radiation from the heat of incoming missiles. However, the technology required for detecting infrared radiation—whether from earthly or celestial sources—did not yet exist. Over the next several decades, as noted in chapter 2, the military invested tens of millions of dollars in research to develop the necessary instruments equipped with infrared detectors (Fig. 3.1).[3] By the early 1960s, these military investments were beginning to pay off. The Air Force, through its Cambridge Research Laboratories (AFCRL), funded the work of physicist Freeman F. Hall, who conducted a sky survey in 1962 at the ITT Federal Laboratory in Sylmar, California, using lead-sulfide (PbS) detectors that operated primarily in the 1- to 3-micron range, just a little longer than optical wavelengths.[4] However, viewing from the ground involved cooling the instruments down so that ambient heat would not saturate the sensors, rendering them unable to detect celestial infrared sources. As a coolant, Hall used frozen carbon dioxide (dry ice). The experiment was a limited success—fewer than 50 sources were identified—although Hall's work

---

2. Fazio interview, 26 May 2009; see also Giovanni G. Fazio, "Planned NASA Space Infrared Astronomy Experiments," *Advances in Space Research* 2, no. 4 (1982): 97–106.

3. Sky survey work in the 1960s was classified; results from this research were first made public in the 1970s and published by Stephan D. Price and Russell G. Walker, *The AFGL Four-Color Infrared Sky Survey: Catalog of Observations at 4.2, 11.0, 19.8, and 27.4 µ*, Publication AFGL-TR-76-0208 (Hanscom AFB, MA: Air Force Geophysics Laboratory [AFGL], Air Force Systems Command, USAF, 1976). For an excellent review of the history of infrared sky surveys, including the pioneering work done by Air Force engineers, see Price, "Infrared Sky Surveys," 2009.

4. Price, "Infrared Sky Surveys," p. 241.

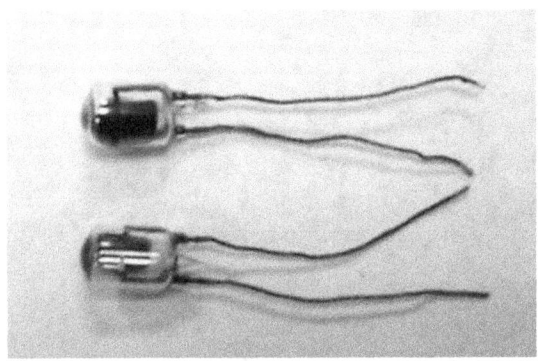

FIGURE 3.1. Lead-sulfide (PbS) detectors (c. 1946) (NASM).

did demonstrate the feasibility of a cryogenically cooled infrared telescope.

The results of Hall's infrared sky survey remained classified, but infrared technology was slowly becoming more available, and a few adventurous scientists were trying it out. Around the same time, a sky survey was being contemplated by physicist Robert Leighton at Caltech.[5] He submitted a funding proposal to NASA in 1962 for the construction of a 62-inch telescope on which to install infrared detectors. Gerry Neugebauer, hired that year as an assistant professor, and several Caltech undergraduate and graduate students helped Leighton to build the telescope. Like Freeman Hall, they used PbS detectors but cooled them with liquid nitrogen, which, at −210°C, was much colder than dry ice. They also used a telescope with a bigger aperture and new filters developed by Optical Coating Laboratory, in Santa Rosa, California.[6] Figure 3.2 (p. 32) shows the completed 12-foot-high telescope. The survey was conducted from January 1965 through the spring of 1968. In his description of an early test, Neugebauer captures the interaction between engineering ingenuity and happy accident that marked the beginnings of infrared research:

> The construction was done here on campus and we erected the telescope ourselves. I remember the first night that we ran with detectors on the telescope.... [I]t was just right outside Bridge [Norman Bridge Laboratory of Physics, on the Caltech campus], in a little alleyway, in which you could only look vertical, essentially. It turned out it was the first night I had gotten the detectors working, and just by luck, we looked up. We didn't know anything about the sky, of course. We looked up and there was a star that looked sort of red, so we pointed toward it. It turns out that it was Beta Pegasus, which is, in fact, one of the three or four brightest stars in the sky at two microns. We just picked that one by sheer accident, because it happened to be overhead in a very narrow range in this alleyway. And everything just worked perfectly.[7]

Because Neugebauer and Leighton were physicists, they had no idea what astronomical results to expect. "I went to the astronomers and asked how many stars will we see? They said 75," Neugebauer remembers. "That was the biggest number I got."[8] Neugebauer and Leighton weren't the only ones to be surprised when their

---

5. Robert B. Leighton, interview by David DeVorkin, Pasadena, CA, 29 July 1977; also Gerry Neugebauer, interview by David DeVorkin, Pasadena, CA, 12 August 1982. Both interviews are from the Space Astronomy Oral History Project, National Air and Space Museum, Smithsonian Institution.

6. For a brief discussion of filters, see Westphal interview, 9 August 1982. Relevant publications of results include Gerry Neugebauer and Robert D. Leighton, *Two Micron Sky Survey: A Preliminary Catalog* (Washington, DC: NASA SP-3047, 1969); and Bruce Murray et al., "Infrared Photometric Mapping of Venus Through the 8–14 Micron Atmospheric Window," *Journal of Geophysical Research* 68 (1963b): 4813–4818.

7. Neugebauer interview, 12 August 1982.

8. Neugebauer interview, 12 August 1982.

FIGURE 3.2. The 62-inch ground-based infrared telescope used for the Caltech Two-Micron Sky Survey (NASM).

Two-Micron Sky Survey, which they published in 1969, gathered light from 20,000 stars.

The Case for Space-Based Observations

In a panel convened by the National Academy of Sciences' Space Science Board in 1965,[9] the discussions included infrared but not as a stand-alone field of science; rather, it was a topic split between the well-established optical and radio groups. Each group was proposing a space-based telescope that would cover some portion (but not all) of the infrared spectrum. Infrared was such a young field that the physicist Frank Low, who had been involved in astronomy for a mere five years, was the token infrared astronomer on both the Working Group on Optical Astronomy (chaired by Lyman Spitzer, the first champion of space-based astronomy) and the Working Group on Radio and Radar Astronomy. As the field of infrared solidified, so would Low's influence. But at that moment, infrared was not a separate field of astronomy, and the role it would play in space science was unclear. Murray and Westphal, Hall, and Leighton and Neugebauer had demonstrated that infrared had promise from the ground. Space-based observations had yet to be proved.

Gerry Neugebauer, however, had had some experience with space-based observations while working on the Mariner 2 program, when he had mapped planetary surfaces in the infrared from space-borne instruments. Mariner, managed at JPL, comprised a series of 10 space probes sent to measure the surface and atmosphere of Mars, Venus, and Mercury. Mariner 2, launched in August 1962, was sent to Venus. Among the instruments on the probe was one by Neugebauer to measure radiation at 8–12 microns in the mid-infrared. Although Neugebauer's instrument worked, he was piggy-backing on another experiment—literally: His infrared instrument was mounted on a microwave dish designed for another experiment. But the dish got stuck. "[We] were just sitting on top of it, and we couldn't point independently," Neugebauer told an interviewer in 1982. "So the microwave lost sync, and instead of making a whole map of Venus as it was supposed to, it made a funny three-angle cut across Venus. So all we got were three swaths across Venus...[whereas] Bruce Murray had been able to make a total map of Venus."[10] Like Murray, Neugebauer was an outsider to the astronomy community—he was a physicist and he studied planets; and, like NASA, he was intrigued with the possibilities of research using infrared detectors.

Neugebauer's instrument had gotten within 22,000 miles of Venus. Technically, it was a major accomplishment; scientifically, it was not. The Mariner team was chagrinned that Bruce Murray, working with Caltech graduate student Robert Wildey and engineer James Westphal, had already done an infrared map of Venus from the ground (see chapter 2). While tens of millions of miles from Venus, Murray had been able to obtain results more useful than those obtained by Neugebauer and the army of engineers and scientists who built and flew the Mariner probe. The results from the infrared experiment on board Mariner 2 exemplified the excitement and disappointment of space-borne astronomy. With attention being focused on infrared observations, some scientists were slowly warming to this new area of astronomy. At the same time, others were lining up against it. While Mariner 2 demonstrated that infrared observations could be made in space, it was less successful in showing

---

9. SSB Study Group, *Space Research: Directions for the Future: The Report of a Study by the Space Science Board*, Woods Hole, MA (Washington, DC: National Academy of Sciences, National Research Council, Publication 1403, 1966).
10. Neugebauer interview, 12 August 1982.

that such efforts were worth it. Many scientists felt that space-based observations were a waste of money—money that could be better spent on ground-based observations. Although there were now more funding sources for astronomy (through NASA, DOD, and NSF), the cost of instrumentation had also increased, and with it, competition among scientists. Moreover, the instruments were custom-tailored to particular wavelengths. Thus, arguments over funding were closely intertwined with issues of scientific merit. One day in 1963, not long after the Mariner 2 mission, Neugebauer happened to be at NASA Headquarters when a letter arrived that openly criticized the space program. Addressed to Senator Clinton Anderson (D-NM), the letter had been written by one of the senator's New Mexico constituents, Bradford A. Smith. What got everyone's attention was that Dr. Smith was a prominent astronomer and Senator Anderson chaired the Senate Committee on Aeronautical and Space Sciences, meaning that he had oversight of both space policy and NASA's budget. Witnessing the arrival of this letter, Neugebauer had the impression that Smith wanted to build up the ground-based program. Certainly, Smith's home state of New Mexico provided excellent sites for viewing the sky. But this was more than a pork-barrel ploy.

Even before Smith wrote his letter, Caltech's Jesse Greenstein had been openly critical of space science programs. Greenstein, who had been disappointed by his experiments on rockets (and who, from his position at Caltech, enjoyed prime access to the coveted and very much ground-based 200-inch Hale telescope), was an influential advisor to NASA and would chair the decadal committee that laid out the future of U.S. astronomy for the 1970s. Lyman Spitzer, who back in 1946 had begun tirelessly advocating for a future that included telescopes in space, and who was working on NASA's Orbiting Astronomical Observatory (OAO) program, recalled in a 1977 interview:

> For years and years, ever since the war, I've been talking about space astronomy to anybody who would be interested enough to listen to me. Jesse Greenstein was a good friend of mine from this period, and he sent up a payload in a V-2 rocket. The only trouble was that the rocket exploded about a hundred feet or so above the launch tower, so all that work was wasted. It sort of soured him on space astronomy for a while! But when I told him I was getting involved in the OAO program, he shook his head, and said, "Well, Lyman, you're young. You'll live to see it fail."[11]

Lyman Spitzer lived for almost 83 years (1914–1997) and is the person for whom the telescope that is the subject of this monograph is named. Spitzer lived long enough not only to see the four OAOs launched but also to see his dream become reality with the launch of the Hubble Space Telescope in 1990. That telescope, the first true observatory in space, might have been named Spitzer instead of Hubble but for NASA's policy of not naming projects after living people. Spitzer would be posthumously honored in 2003, when SIRTF was renamed the Spitzer Space Telescope—after it was determined that the instruments were working.

NASA Loses Luster

Just as 1967 brought much attention to infrared astronomy, another event was bringing attention

---

11. Lyman Spitzer, interview by David H. DeVorkin, Pasadena, CA, 8 April 1977, Niels Bohr Library & Archives, American Institute of Physics, College Park, MD, *https://www.aip.org/history-programs/niels-bohr-library/oral-histories/4901-1* (accessed 30 August 2016).

to NASA—unwelcome attention. While scientists were criticizing NASA's budget priorities, the Agency found itself under intense scrutiny by both Congress and the media after a fire during a launch simulation claimed the lives of three Apollo astronauts. As NASA's Associate Administrator at the time, Homer Newell, recalls:

> Under the best of circumstances the Apollo 204 fire on 27 January 1967 would have been difficult to live down. But coming at a time when the country was becoming more concerned about a variety of problems other than whether the United States was or was not ahead of the Soviets in space, the impact of the accident upon the [A]gency was immeasurably increased. A great deal of [NASA] Administrator Webb's time was taken up in recouping for NASA the respect it had been building up in the Mercury, Gemini, and other programs, and in regaining the confidence of the Congress. That in Apollo the United States was on trial, as it were, before the whole world had much to do with the program's continuing to receive support. But in the aftermath of the congressional hearings and internal NASA reviews, Webb began to sense a slackening of support for the space program.[12]

According to Newell, this lack of support did not deter Thomas O. Paine, who became NASA Administrator in March 1969, shortly after the congressional hearings on the Apollo fire ended and Webb retired. By 1969, with funding for Apollo winding down, NASA was looking for new ways to contribute to national policy. After the Moon landing, some felt that an equally grand project was needed. Reflecting those opinions and hopes (and not the economic realities), Paine advocated several big projects: a lunar base, a space station, and a reusable space transportation system—the Space Shuttle.[13]

To get political support for development of the Shuttle, planners wanted to show that there was widespread support and demand for regular crewed spaceflights. By including onboard experiments (something that was not central to the Apollo Moon missions), they hoped to make support for a Shuttle more compelling to both scientists and Congress. Likewise, a space station was meant to appeal to scientists and compete with the Soviets, who were moving forward with the first orbiting space platform, Salyut (launched in 1971).[14]

To guide NASA and Congress, the scientific community was called upon to develop a list of priorities for policy-makers.[15] Under the sponsorship of the National Academy of Sciences, which

---

12. Homer E. Newell, *Beyond the Atmosphere: Early Years of Space Science* (Washington, DC: NASA SP-4211, 2004), pp. 284–285, available at *http://history.nasa.gov/SP-4211/cover.htm* (accessed 30 August 2016).

13. Ibid.

14. For a discussion of the scientific community's concerns with science being subservient to engineering and piloted mission goals of NASA throughout the 1960s, please see Tatarewicz, pp. 103, 136–137; and Walter A. McDougall, *... the Heavens and the Earth: A Political History of the Space Age* (Baltimore: Johns Hopkins University Press, 1985), esp. pp. 389–402. To garner support for the Space Shuttle, panels of scientists were tasked with finding payloads that would warrant reusable and long-term space platforms; see, for example, *Proceedings of the Space Shuttle Sortie Workshop, Greenbelt, MD, Volume 1: Policy and System Characteristics and Volume II: Working Group Reports* (Greenbelt, MD: NASA Goddard Space Flight Center, 1972); and *Final Report of the Space Shuttle Payload Planning Working Groups, Volume 1: Astronomy*, Publication NTRS 1974007405 (Greenbelt, MD: NASA Goddard Space Flight Center, 1973); while Fred Witteborn, who conceived SIRTF, served on the panel that issued the report *Scientific Uses of the Space Shuttle* (Washington, DC: National Academy of Sciences, National Research Council, 1974).

15. Astronomy Survey Committee (Jesse Greenstein, chair), *Astronomy and Astrophysics for the 1970s, Volume 1: Report of the Astronomy Survey Committee and Astronomy and Astrophysics for the 1970s, Volume 2: Reports of the Panels* (Washington, DC: National Academy of Sciences, 1972), hereafter cited as "Greenstein report."

had been advising Congress and presidents on policy since Abraham Lincoln was in office, scientists from many fields gathered between July 1969 and October 1971 to identify the most promising research directions and determine the best use of funding for ground-based, airborne, and space-based experiments. Jesse Greenstein chaired the Astronomy Survey Committee, which brought together scientists studying all regions of the electromagnetic spectrum and asked them to prioritize research programs within their respective domains. The infrared panel included John Gaustad (UC-Berkeley) as chair, Eric Becklin (Caltech), Fred Gillett (UC-San Diego), James Houck (Cornell), Harold Larson (Arizona), Robert Leighton (Caltech), Frank Low (Arizona), Douglas McNutt (NRL), Russell Walker (AFCRL), and Neville Woolf (Minnesota).

The infrared panel recommended a sky survey to obtain an overall picture, to be followed by a cooled infrared space telescope of great sensitivity. The cost estimate for the proposed infrared telescope was $100 million, which was 20 times more expensive than anything else the infrared panel recommended. The panel submitted its recommendations to Greenstein's committee, which was charged with synthesizing the various panel reports. The committee's goal was to provide NASA and other policy-makers with a consensus on scientific research priorities that had been weighed against economic considerations. The resulting decadal survey for the 1970s, also known as the Greenstein report, ultimately recommended that highest priority be given to solar, optical, and x-ray programs—areas that had already produced important findings and could be counted on to deliver solid research results, even within a limited national budget. Regarding the cooled infrared space telescope, Greenstein's report was dismissive: "This instrument will be needed in time, but technological problems are severe, cost is high, and much can be done from a stratospheric platform first [i.e., from balloons, rockets, and aircraft]."[16]

As the 1960s drew to a close, the economic, political, technical, and scientific forces that had given rise to infrared astronomy were pulling it in different directions. As a scientific field, infrared astronomy did not have a central research question or a way to unite all the stakeholders in a common pursuit. Universities were vying for leadership in this new scientific field, but most of the required technology did not yet exist. Aerospace companies were competing for contracts to develop that technology. Although NASA was encouraging astronomers to propose experiments as a way to increase the scientific support and content for the Shuttle Program—and thus its political appeal to Congress—scientists remained doubtful about NASA's commitment to space-based science. Furthermore, it seemed clear that national priorities would now focus on cutting domestic spending and ending the Vietnam War, as Nixon had indicated in his 1968 election campaign. And with the White House turning its lights off at night due to the 1970s energy crisis, the economic mood on funding lavish space programs had grown dark.

Infrared Up in the Air

It was in this chilly political climate that Fred C. Witteborn conceived of building a cryogenically cooled infrared space telescope. Witteborn was the chief of the new Astrophysics Branch at NASA's Ames Research Center. Ames had been founded in 1939 by the U.S. government to conduct aeronautical research, as a backup facility for the Langley Research Center, in Virginia, in case

---

16. Greenstein report, vol. 2, p. 93.

of a wartime attack on the East Coast. Ames is at the southern end of San Francisco Bay and lies off California Highway 101, halfway along the 16-mile stretch between Stanford University to the north and Silicon Valley to the south. Both geographically and intellectually, Ames is situated between the worlds of scientific research and technology development. Witteborn had to enlist both these realms to produce his cryogenically cooled infrared space telescope.

Witteborn had come to Ames for a summer job in 1957 while working on his doctorate in physics at Stanford. His graduate advisor, William Fairbank, was a low-temperature physicist who had started a research program to test part of Albert Einstein's general theory of relativity.[17] Key to this research was finding a way to slow matter down by cryogenically cooling the particles. "P. W. Selzer, W. M. Fairbank, and C. W. F. Everitt were the inventors of a vent valve that enabled the storage of superfluid helium in space," Witteborn recently recalled. "This solved the trouble with cryogens: When you have zero gravity, what's to keep them in the bottom of the tank? But if it's superfluid with a proper temperature gradient across the porous plug vent, you're able to contain superfluid."[18] Witteborn learned about these techniques while working on his dissertation, which he completed in May 1965.

Witteborn continued to work at Ames while pursuing his doctoral degree. During this time, his boss at Ames, Michel Bader, started the airborne astronomy program by modifying commercial aircraft. Even though Ames did not have any astronomers on staff, it was renowned for its aerospace development capabilities. It has its own airport, Moffett Field, and the world's largest wind tunnel—so large it can accommodate a Boeing 727. Airborne astronomy was a perfect way to leverage Ames's engineering strengths in order to move the Center beyond aeronautics design and toward astrophysics research (while further validating its identity as a part of NASA). Airborne astronomy was economically attractive—cheaper than using rockets, applicable at all wavelengths, more flexible than a mountaintop location, and updatable as technology improved. In 1964, Bader had secured a Convair 990 airplane on which to conduct scientific experiments in "sortie mode," a single flight to achieve a narrowly defined mission, in which the necessary equipment is carried aloft, measurements are made, and then brought back to Earth. Although NASA owned the plane, it was made available to scientists from around the world. Among the first to design airborne experiments were Gerald Kuiper, Dale Cruikshank, and Frank Low from the University of Arizona and Gerry Neugebauer from Caltech.[19]

The infrared telescope that Frank Low configured for the Convair 990 made it possible to take observations at 40,000 feet while neutralizing much of the interference from the atmosphere; observations were limited, however, by the size of the aircraft's 12-inch windows and 1-inch-thick

---

17. Incidentally, this research program evolved into the NASA-Stanford Gravity Probe-B project; for an early discussion, see C. W. Francis Everitt, William M. Fairbank, and Leonard I. Schiff, "Theoretical Background and Present Status of the Stanford Relativity-Gyroscope Experiment" in *The Significance of Space Research for Fundamental Physics:* Proceedings of the Colloquium of the European Space Research Organization, held 4 September 1969 at Interlaken, Switzerland (ESRO SP, No. 52): p. 33–43 (proceedings available at *http://www.iaea.org/inis/collection/NCLCollectionStore/_Public/03/024/3024818.pdf*).

18. Fred C. Witteborn, interview by author, Mountain View, CA, 2 September 2008.

19. Research publications that resulted from work on the Convair 990 aircraft included G. Münch et al., "Infrared Coronal Lines. II. Observation of [Si x] $\lambda 1.43\ \mu$ and [Mg VIII] $\lambda 3.03\ \mu$," *Astrophysical Journal* 149 (1967): 681–686; also see Gillespie, Carl, ed., *Gerard P. Kuiper Airborne Observatory and Learjet Observatory Plus Astronomy-Related Publications from the NASA Convair 990 Aircraft* (Moffett Field, CA: Ames Research Center, 1991).

glass.[20] Low was soon developing a telescope to work with an open port—basically a hole in the side of the plane—initially under sponsorship by AFCRL and later by NASA.[21] The instrument (shown in Fig. 3.3) was installed on a Learjet, and NASA gave Ames responsibility for conducting the flights.

"One of the things that we were supposed to do when the Astrophysics Branch was first born was to fly on the Learjet with a small airborne telescope," Witteborn reports; "Headquarters had twisted [Low's] arm to make it available to the rest of the IR [infrared] community. We had the misfortune of being made the host center for his telescope to fly on a NASA airplane. This created some friction."[22] Low had a strong personality, and by all accounts he was a forceful if gruff advocate for infrared research. Along with demanding personalities, researchers on the Learjet had to cope with a physically demanding environment. Astronomers and experimental astrophysicists are used to cold, dark observing conditions, but the Learjet was in a different league. Even though the cabin was partially pressurized and heated, the air remained so cold and thin that everyone on board wore oxygen masks and warm clothes.[23] They flew sorties at 2:00 or 3:00 in the morning and always needed scientists who were willing to give up their sleep to run the equipment. "I lost the need to have regular hours after that project," Witteborn said.

"I never worried again about when I got up. But [Lawrence J.] Caroff was in the [Ames] theoretical branch. So for him to volunteer was something of a sacrifice, but he was a very good telescope operator." Although Caroff was one of the only theoreticians who flew, the Learjet program benefited many experimentalists besides Low. According to Witteborn, "we built our own instruments and flew on the Learjet and did some early infrared research, spectroscopic research. There had been some research done on the [Convair] 990, but this was the first open-port spectroscopy done from the Learjet with our instruments."[24]

The airborne astronomy program at Ames grew throughout the 1960s, and by 1971 it was reorganized to bring together all of the infrared science projects. As there was already a NASA "Infrared Branch" at Goddard Space Flight Center (and fierce competition between the two Centers), Ames called their new infrared division the Astrophysics Branch. Bader asked Witteborn to lead the new branch and to find a way to continue hosting astronomical missions. "When Mike Bader heard that they wanted science instruments for the sortie mode on the Space Shuttle, he said, 'Gee, you guys ought to put a telescope on there.' He had an entirely different idea than I of what it should be," Witteborn remembers. "I realized right away that for infrared we needed a cryogenic telescope."[25]

---

20. Wendy Whiting Dolci, "Milestones in Airborne Astronomy: From the 1920's to the Present," *AIAA, 1997 World Aviation Congress, October 13–16, 1997,* Anaheim, CA. This article is available in print as NASA ARC 975609 and in digital format at *https://www.sofia.usra.edu/sites/default/files/97-Whiting_AeroHistory.pdf* (accessed 30 August 2016).

21. Price, "Infrared Sky Surveys," pp. 248–249.

22. Witteborn interview, 2 September 2008.

23. NASA Ames Research Center, "Lear-Jet Airborne Observatory Investigators Handbook" (Moffett Field, CA: NASA TM-108623, 1974).

24. Witteborn interview, 2 September 2008.

25. Witteborn interview, 2 September 2008. In regard to the Space Shuttle, sorties were discussed at least as early as September 1971; see Judy A. Rumerman, ed., *NASA Historical Data Books, Volume V: NASA Launch Systems, Space Transportation, Human Spaceflight, and Space Science, 1979–1988* (Washington, DC: NASA SP-4012, 1999), p. 462. The chapter on space science is available at *http://history.nasa.gov/SP-4012/vol5/vol_v_ch_4.pdf* (accessed 30 August 2016).

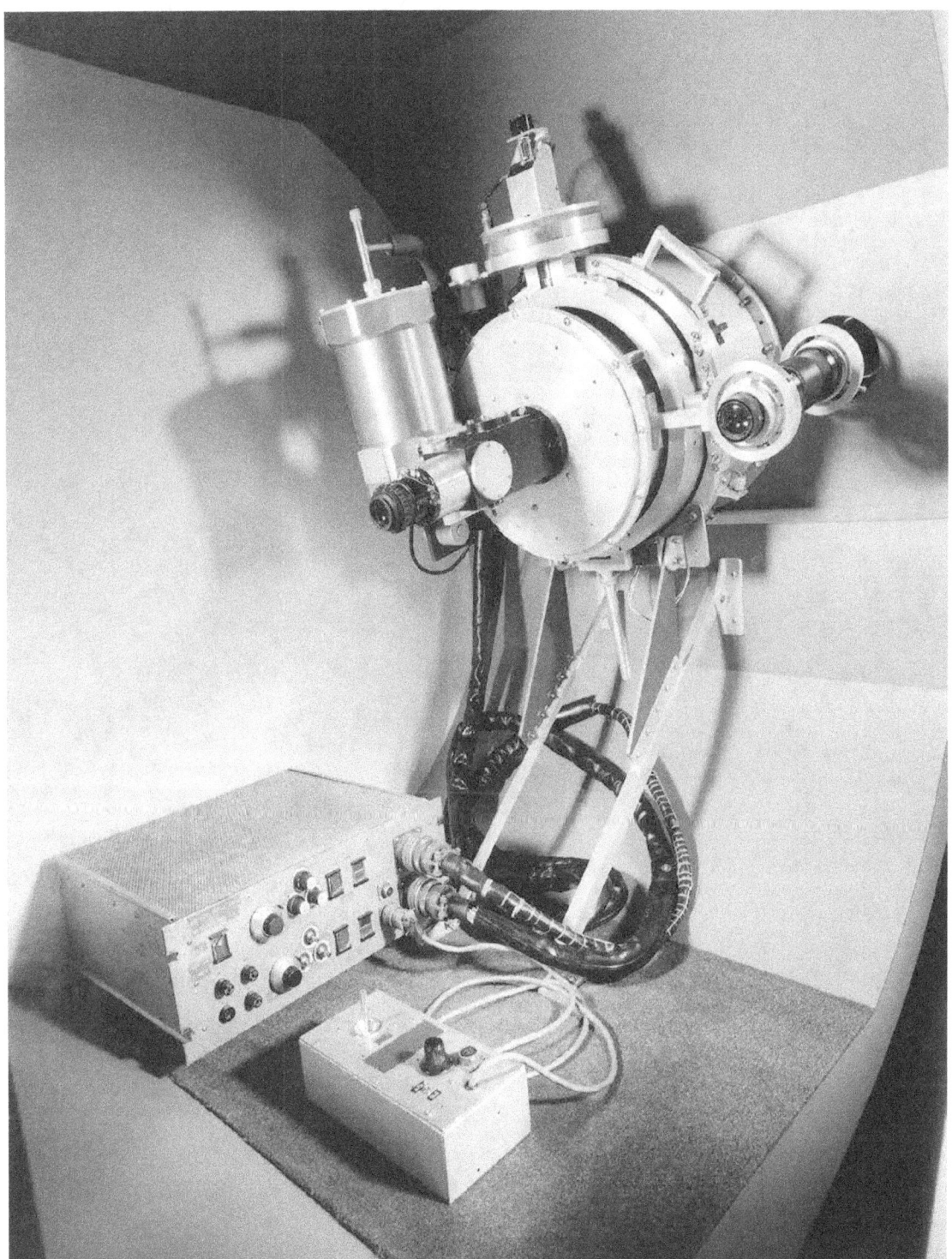

FIGURE 3.3. Telescope with cryogenically cooled infrared instruments (c. 1972), developed by Frank Low and flown aboard the NASA Ames Learjet (NASM).

CHAPTER 4

SIRTF as a Shuttle-Based Infrared Telescope

To sketch out what such a telescope might look like, Witteborn assembled a team at Ames that included Lou Young, a seasoned engineering manager from the Apollo program; Larry Caroff, the airborne-telescope-operating theoretician; and Eric Becklin, co-discoverer of the Becklin-Neugebauer object and an active participant in the Ames airborne astronomy program after earning his doctorate from Caltech. Their resulting design (Fig. 4.1) was originally conceived as a cooled telescope with a 1-meter mirror. The technical instruments envisioned as early as 1973 (Fig. 4.2, p. 42) hint at the photometric and spectroscopic instruments that were actually flown 30 years later.

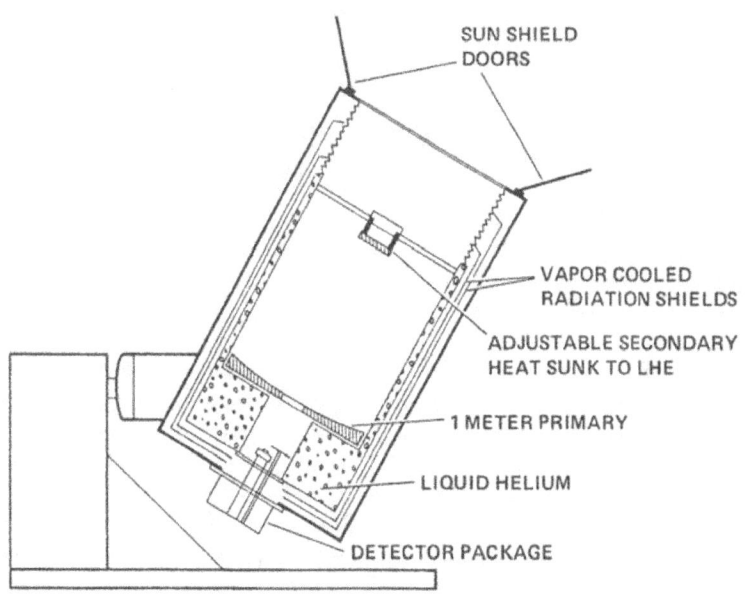

FIGURE 4.1. Design for SIRTF as a Shuttle-attached payload, by Fred Witteborn and colleagues at Ames (c. 1971).

> The 1.5-m infrared telescope (Figure 2) is envisioned as a Cassegrain with an f/2 primary possessing at least a 0.5° field of view, a modulating secondary mirror, cooled baffles, a movable sun shield and a removable thin plastic window for protection from contaminants during early phases of a mission. The telescope would be mounted on gimbals (possibly on a swing table) on a Sortie pallet and would constitute approximately 1/2 of a Sortie payload. Manned access to the focal plane would not be necessary. A rotatable tertiary mirror in the instrument bay would be commanded to direct the light beam into any of several different instruments. Instruments could be interchanged from mission to mission to accommodate the needs of different investigators. A complement of instruments and detectors might include:
>
> - A broad-band infrared filter photometer using a liquid helium-cooled detector.
> - An infrared photoconductor detector array (LHe-cooled doped germanium) for flux measurements with spatial resolution.
> - A Fourier spectrometer (LHe-cooled) for medium resolution (0.1 cm^{-1}) infrared spectroscopy.
> - An infrared polarimeter (both linear and circular)
> - A grating spectrometer with multichannel detectors for intermediate band infrared spectrophotometry.
>
> Rotatable disks between the tertiary and secondary mirrors would contain an assortment of filters and an optional beam chopper.

FIGURE 4.2. Specification of possible SIRTF design and instrument suite (c. 1973) (NASA Goddard Space Flight Center).

Witteborn represented Ames at a conference on sortie-mode astronomy held at Goddard Space Flight Center (GSFC) in Greenbelt, Maryland, from 31 July to 4 August 1972. This conference, co-chaired by Larry Caroff, was the first step in sortie-mode planning for the Space Shuttle. "Bill [William F.] Hoffmann was there and he gave a paper on his balloon-borne, infrared telescope," Witteborn said. "He and I got together and discussed the importance of having a cold telescope on the [S]huttle."[1] The "cooled infrared telescope," as it was then called, would become SIRTF and would eventually be renamed the Spitzer Space Telescope.

While the acronym SIRTF remained constant until December 2003 (when the telescope's name was changed to Spitzer, four months after launch), the meaning of the acronym did change. Initially, SIRTF stood for Shuttle Infrared Telescope Facility because SIRTF was originally intended to be flown as an attached payload to the Space Shuttle. Choosing the Shuttle as the launch vehicle for SIRTF solved some problems but created others. If rockets were capable of sending humans to the Moon, why commit to launching SIRTF

---

1. Witteborn interview, 2 September 2008. See also *Proceedings of the Space Shuttle Sortie Workshop, July 31–August 4, 1972, Volume 1: Policy and System Characteristics and Volume 2: Working Group Reports* (Greenbelt, MD: NASA Goddard Space Flight Center, 1972).

on a system that had not yet flown? As Witteborn recalls, "We were told by NASA Headquarters in no uncertain terms that if you are going to get NASA's money to fly scientific payload missions in space, you had better plan on doing it on the [S]huttle... [which] was going to be so much cheaper. They were going to fly payloads for $100 per pound."[2] Rocket-borne payloads cost $1,000 per pound. Moreover, NASA was discontinuing manufacture of expendable rockets and therefore had a limited stockpile with many demands on it—from launching telecommunications satellites to carrying military payloads.[3]

The 1972 conference on sortie-mode astronomy at Goddard brought together members of NASA from Headquarters, Marshall Space Flight Center (MSFC), Ames Research Center (ARC), and the Jet Propulsion Laboratory (JPL). Working groups were assigned to different subfields of astronomy, including solar, x-ray, optical, and high-energy physics. The Working Group on Infrared Astronomy was co-chaired by Caroff and Maurice Dubin, from Headquarters. Participants included, besides Witteborn, William Hoffmann (University of Arizona), Theodore P. Stecher (GSFC), Reinhard Beer (JPL), and Thomas Wdowiak (MSFC). Also representing Headquarters was Nancy Boggess.

Boggess was a senior staff scientist at Headquarters; she joined NASA after receiving her doctorate in astrophysics from the University of Michigan in 1967. Her boss was Nancy Roman, who was also present at this meeting and busy chairing the Optical Astronomy Working Group, which was discussing the Hubble Space Telescope project. As the scope of NASA's involvement in astronomy grew, it became necessary for Roman to delegate the management of various subfields; Roman put Boggess in charge of the new infrared program. As Boggess later recalled:

> At that point, infrared was just coming into view at NASA. Nancy Roman supported several infrared researchers at Caltech and at the University of Minnesota, so she had her hands full and asked me to work on that.... Since nobody at Headquarters knew much about infrared, she gave me the opportunity to go around to the various universities that we were supporting and learn a lot of infrared astronomy from the researchers. I'll be indebted to her for it, because it helped me manage the program much better having that background.... It really gave me a good fundamental understanding of what the problems were in the infrared and how to overcome them and what the science goals could be in the infrared.[4]

Following the workshop at Goddard, discussions about SIRTF continued. Roman chaired a subcommittee within the Payload Planning Working Group, an ad hoc committee of researchers from NASA Centers and scientists from the United States and abroad to oversee the scientific uses of the Space Shuttle.[5] Other subgroups met to focus on possible Shuttle payloads for solar, high-energy, and atmospheric science; whatever wasn't covered by those categories was put under the aegis of Roman's group, the Astronomy Science Working Group, which met from November 1972 to April 1973 and included

---

2. Witteborn interview, 2 September 2008.

3. President Reagan reinstated the expendable launch vehicle program after the loss of Challenger; see Marcia S. Smith, *Space Launch Vehicles: Government Activities, Commercial Competition, and Satellite Exports*, Congressional Research Service, Library of Congress (IB93062 CRS Issue Brief for Congress), updated 3 February 2003.

4. Nancy Boggess, interview by author, Boulder, CO, 19 March 2009.

5. *Final Report of the Space Shuttle Payload Planning Working Groups, vol. 1*. cf. n. 6.

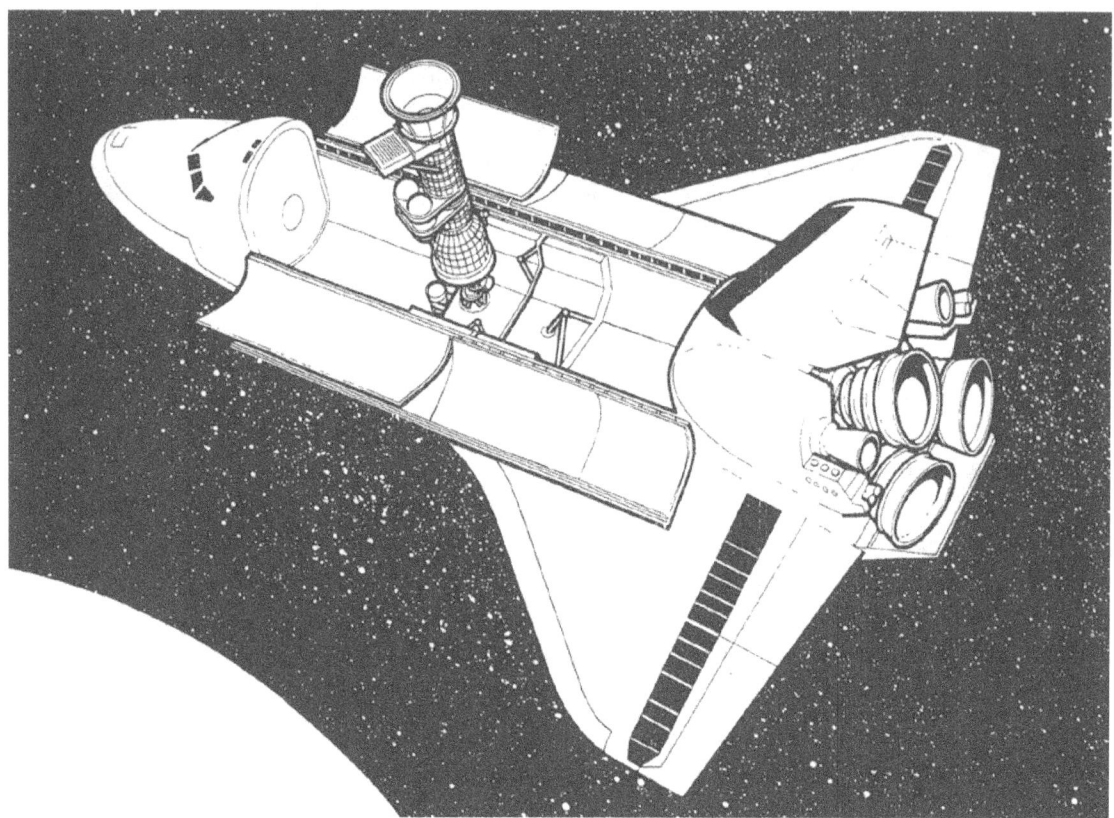

FIGURE 4.3. SIRTF as a Shuttle-attached payload (c. 1980) (NASA).

scientists from several NASA Centers (Goddard, Ames, Johnson, Marshall, and Headquarters), from the NRL, and from many of the major ground-based observatories, including Yerkes, Palomar, Lowell, Washburn, Lick, Steward, and McDonald in the United States and observatories in Marseille and Paris.

The initial idea for SIRTF was a cryogenically cooled infrared telescope mounted to a reusable instrument pallet. Observation would begin when the Shuttle achieved its low-Earth orbit and opened its cargo-bay doors to the sky (Fig. 4.3).[6]

The telescope would be cooled to below 20 K and have a mirror of 1.0–1.5 meters, its size being restricted to what could fit in the Shuttle's bay (15 feet in diameter and 60 feet long). Since the initial 1971 design, the telescope had nearly tripled in length and was now over 24 feet long.

It was expected that scientists would use SIRTF to conduct sky surveys as well as to gather data on galactic composition, early conditions of the universe, and dust clouds. To meet these science objectives, the original design of SIRTF relied on two types of measurements:

---

6. This Shuttle configuration is largely the same as one presented in Fred C. Witteborn and Lou S. Young, "Spacelab Infrared Telescope Facility (SIRTF)," *Journal of Spacecraft and Rockets* 13, no. 11 (November 1976): 667–674, esp. p. 670. Witteborn and Young's article is based on an earlier conference paper, "A Cooled Infrared Telescope for the Space Shuttle: The Spacelab Infrared Telescope Facility," that was published in January 1976 as AIAA paper # 76-174; see also the presentation by Lou Young to NASA on the Statement of Work, Phase A, for SIRTF (Item PP05.04, Larry A. Manning Papers 1967–1988, Box 2, Folder 3, housed in the NASA Ames History Office, NASA Ames Research Center, Moffett Field, CA).

photometric and spectroscopic. Photometry shows an object's energy output and temperature, whereas spectroscopy shows its chemical composition.

The Astronomy Science Working Group expressed concerns about contamination from the Shuttle environment. Dust, moisture, and fumes could all cause light to scatter before it could be read by the instruments—or, worse, these substances could stick to the telescope's mirror and obscure the data entirely. Except for an occasional cleaning of the mirrors by a Shuttle astronaut, the telescope required no maintenance. To prevent contamination, the working group proposed that SIRTF be a free-flying satellite. Instead of being operated on board the Shuttle, SIRTF would be launched by the Shuttle and placed in orbit using a space tug called an orbital maneuvering vehicle. SIRTF would then be periodically visited and serviced by Shuttle crews to maintain the instruments and replace the cryogen. However, this was a bit premature, as neither the Shuttle nor the space tug existed yet.

In its final report, Roman's Astronomy Science Working Group recommended that various Shuttle payloads include three infrared

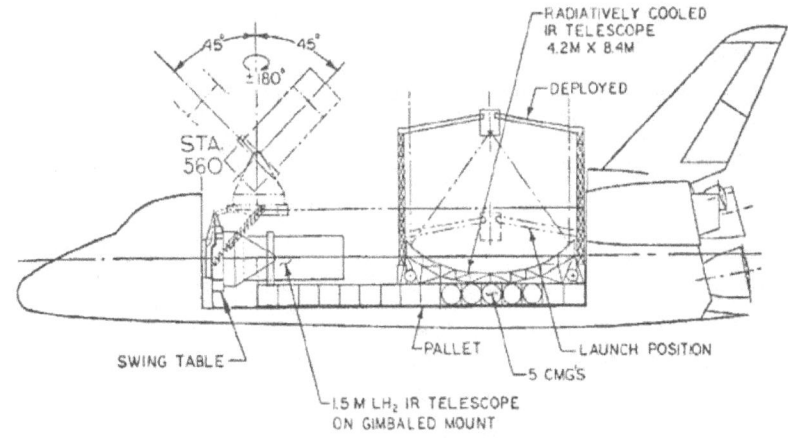

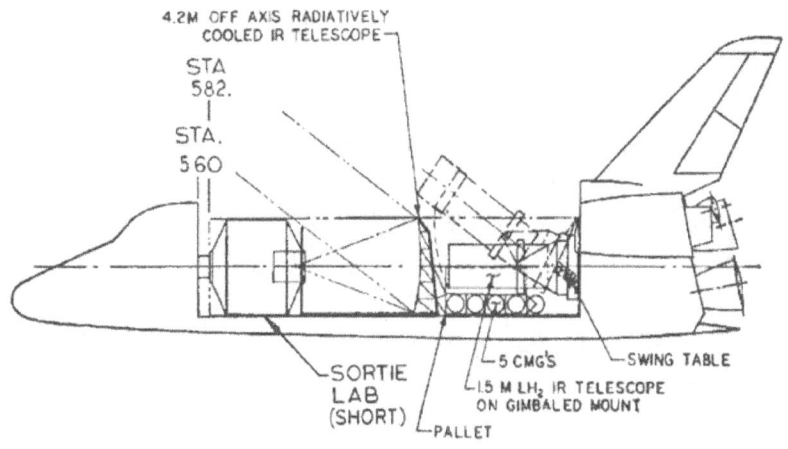

FIGURE 4.4. Proposed Shuttle payloads with multiple infrared telescopes (*Final Report of the Payload Planning Working Groups, vol. 1: Astronomy*).

telescopes, as well as the Hubble Space Telescope (primarily for optical wavelengths), which was already a few years into planning. The first project in the infrared would be a small cooled telescope, which eventually developed into SIRTF.[7] The second infrared project was to be a pair of large, uncooled telescopes, the first of which would have a mirror 4 meters in diameter and the second a mirror 8 meters in diameter, both operating at around 200 K, for gathering far-infrared

---

7. *Final Report of the Payload Planning Working Groups, vol. 1: Astronomy* (Greenbelt, MD: NASA Goddard Space Flight Center, 1973).

observations at more than 100 microns.[8] The working group speculated that, in time, the small cooled infrared telescope would be flown with one of these larger uncooled telescopes (see Fig. 4.4, p. 45).

The next step in the development of SIRTF was to broaden discussions to include scientists involved in setting national policy. In July 1973, the NAS's Space Science Board (SSB) met in Woods Hole, Massachusetts, to discuss the scientific uses of the Space Shuttle. At this meeting, 91 scientists, including 13 from outside the United States, convened to consider how the Space Shuttle might be used for research in atmospheric, planetary, and life sciences; high-energy physics; and optical and infrared astronomy. The infrared astronomy panel was chaired by Bill Hoffmann (University of Arizona) and included Rudolf Hanel (GSFC), David Rank (Lick Observatory), and Fred Witteborn (ARC), all of whom had participated in Roman's Astronomy Science Working Group. Joining them on the panel were Richard Jennings (University College London), Gerry Neugebauer (Caltech), and Stephan Price (AFCRL).[9]

The top recommendation by the SSB's infrared panel was for a 1.0- to 1.5-meter cryogenically cooled telescope, to be launched in 1981 in sortie mode.[10] This proposal was consistent with the earlier recommendations of Roman's Astronomy Science Working Group; however, it contradicted the 1972 Greenstein report, the National Academy of Sciences' decadal survey for the 1970s, which had recommended that infrared research be conducted on the ground and in the stratosphere only, while x-ray, optical, and gamma-ray research should be done in space.[11] The inconsistency was surprising, in that the SSB was an NAS subcommittee. The divergence from Greenstein's decadal survey was due to the political positioning of the Space Shuttle Program. While the cooled infrared telescope was not a priority within the wider scientific community, it was a priority for those trying to generate support for the Shuttle by identifying possible scientific payloads to carry.

SIRTF was to be followed by other telescopes—some that were smaller or designed for a specific experiment, such as a sky survey, and culminating in a multipurpose telescope facility with a 10-meter mirror, to be assembled in space via the Shuttle. Table 4.1 shows the proposed schedule of Shuttle-borne infrared experiments, as envisioned by the SSB panel.

While these recommendations were intended to increase support for the Space Shuttle, particularly among the scientific community, they were somewhat disconnected from those promoted by John E. Naugle, NASA's Associate Administrator for Space Science and Applications. In Naugle's testimony to the U.S. Senate on the use of the Space Shuttle and in the popular press, the Space Shuttle had been sold mainly on the basis of its ability to economically gather data that addressed problems on Earth, such as tracking weather and crops.[12] Infrared remote sensing (which focuses on Earth) was an important part of this solution, whereas infrared astronomy (which focuses on deep space) was not. Moreover, the U.S. government was then dealing with the 1973 OPEC

---

8. An incarnation of the 4-meter telescope that was proposed in 1973 became part of the Herschel mission, a joint program of the European Space Agency and NASA that launched 14 May 2009.

9. National Research Council, *Scientific Uses of the Space Shuttle* (Washington, DC: The National Academies Press, 1974).

10. National Research Council, *Scientific Uses of the Space Shuttle*, p. 99.

11. Greenstein report, vol. 1 (1972), pp. 11, 85, and 91.

12. *Space Shuttle Payloads: Hearing Before the U. S. Senate Committee on Aeronautical and Space Sciences*, 30–31 October 1973, Washington, DC; John E. Naugle, "Research with the Space Shuttle," *Physics Today* 26, no. 11 (1973): 30–37.

TABLE 4.1. SIRTF as a 1-meter telescope to be flown on the Space Shuttle in 1981.

	Year												
	'79	'80	'81	'82	'83	'84	'85	'86	'87	'88	'89	'90	'91
Sortie													
LH$_2$-cooled, 1 m			½	2 × ½	2 × ½	2 × ½	2 × ½	½	½				
Ambient-temperature, 1 m[a]	¼	2 × ¼	¼	2 × ¼	2 × ¼	2 × ¼	2 × ¼						
Ambient-temperature, 3 m[b]					1	1	1	1	1	1			
Interferometer 10-m baseline								½	½				
LH$_2$-cooled 2.5 m[b]									1	2	2	2	2
Interferometer, 1.5-km baseline[b]										½	½	½	½
Ambient-temperature, 10 m[c]							1				1	¼	¼
Small Free-Flyers and Rocket Payloads													
IR survey			●			●							
Submillimeter background					●			●					
Millimeter background				●			●						
IR monitors/Explorers						●			●	●	●	●	●

a Operation of first flights will be during test phase of Shuttle.
b Whether these instruments are sortie mode or free-flyers should be decided on basis of early sortie results.
c To be assembled in space during sortie mission and then set free.

oil embargo and the resulting economic crisis. Although NASA had won congressional approval to move ahead with the Shuttle, infrared astronomy's budget within NASA was expected to be small—infrared research was to be allocated $2 million a year, or 2.6 percent of the Office of Space Science and Applications (OSSA) astronomy budget, throughout the 1970s. In 1972, the National Academy of Sciences recommended doubling the funds for infrared to $4 million a year; none of that money was earmarked for an infrared space telescope.[13]

Assessing the Shuttle Environment

SIRTF had no priority in NASA's overall science budget; however, funds for science could be had from the Shuttle program. That it was politically and economically desirable to put an infrared telescope on the Space Shuttle was a weak argument; weaker still was the notion that the Shuttle would provide observing conditions that could satisfy scientific and technical needs.

The major technical problem with having an infrared telescope on the Shuttle was contamination. One source would be the debris and

---

13. Greenstein report (1972), vol. 1, p. 11; and vol. 2, p. 380.

outgassing of materials from the Shuttle. Another source would be the astronauts, who would create vapor clouds by simply exhaling and vibrations by moving about. For a highly sensitive telescope, the moisture and jitter on the Shuttle might be worse than anything encountered on the ground. "As the science teams began to look at it, they saw that contamination was an unknown," Witteborn said.[14] They estimated that infrared sensors would pick up particles as small as 2 microns up to 10 kilometers away. To put the problem in perspective, the period at the end of this sentence measures about 400 microns. Everyone was right to be worried; conclusive evidence that the Shuttle environment gave off too much heat and vapor for infrared observations would come in 1985 from Giovanni Fazio's IRT, which took measurements on board an orbiting Shuttle.

Even if the Shuttle were contaminant-free, a major scientific concern remained as to the amount of viewing time available on a Shuttle-borne telescope. With astronauts on board, the Shuttle could stay in orbit for only a few weeks. The scientists wanted a month or more of observing time. As Witteborn recalls:

> [NASA told us to] count on a week or two [in orbit, with] 60 [opportunities to launch] a year. Then, if your instrument didn't work out quite the way you wanted, you would have a chance to fix it between missions. Well, it sounded very similar to what Ames was doing with the 990 [the Convair aircraft that replaced the Learjet]. You build an instrument, you fly it, you come down again, and you change it. Also, on the Kuiper Airborne Observatory [the C-141 aircraft that began operations in 1974]—again, you fly your instrument for eight to 10 hours, and then you have a chance to put new cryogen in it, fix it up, and send it up again. So the idea was appealing.[15]

Nevertheless, Witteborn and the rest of the panelists at the 1973 meeting at Woods Hole made it clear in their report that they preferred a longer mission. Despite the imperative to use the Shuttle, they had expressed a preference for a free flyer as early as 1972. "We really much preferred to have our own missions as a separate satellite," Witteborn said.[16] If disconnected from the Shuttle, SIRTF could then be left in orbit, lengthening the observational periods and reducing the possibility of contamination. But in 1972 no one knew for certain what the real effects of a Shuttle environment were, as the Shuttle itself was still in the early stages of design.[17]

To understand the anticipated operational environment of the Shuttle, NASA provided money to conduct engineering studies and brought together a group of experts to oversee the process. "Study money was made available," Witteborn said. "It was a whole $25,000."[18] The slim budgets of the 1970s intensified the competition among astronomers over funding and project priority. Witteborn continues:

> A certain person in the astronomical community was intensely outraged over this—"All that money was being squandered at Ames for this project." So it was reexamined and NASA Headquarters decided we could

---

14. Witteborn interview, 2 September 2008; see also Witteborn and Young, "SIRTF" (1976).

15. Witteborn interview, 2 September 2008.

16. Ibid.

17. For preference of scientists to have the telescope *remain* in orbit, see *Proceedings of the Space Shuttle Sortie Workshop*, vol. II, p. 15-2.

18. Witteborn interview, 2 September 2008.

do a study of the contamination environment of the Space Shuttle and its effect on infrared astronomical observations. Martin Marietta won this contract, which in hindsight appeared to be a traditional buy-in. A big company like that couldn't do much for $25,000, but they did an impressive job. They reported on how you measure the contamination looking at the materials and how they outgas. They obviously had done a study before on this and gave us the same stuff again, but applied to the Shuttle. It was to their advantage to do it, too. And we asked for more. We asked for a literature search on infrared techniques for space astronomy.[19]

The Martin Marietta report was the first technical study of what would develop into SIRTF.

Developing a Design for SIRTF

Not long after the contamination study of the Shuttle environment, NASA put out a call for proposals to develop a design for this new cooled infrared Shuttle telescope. Martin Marietta was interested in responding to the call and hired several astrophysicists to review its proposal. One of those consultants was Robert D. Gehrz, who had recently obtained a doctorate in physics (1971) from the University of Minnesota. His dissertation was on astrophysics and involved the construction of infrared detectors used on ground-based telescopes. Shortly after joining the faculty at the University of Wyoming in the fall of 1972, Gehrz reported,

I was approached by Martin Marietta Aerospace to consult with them on a proposal they were writing to do the initial design studies for SIRTF. At the same time, they hired Jim Houck and George Rieke as consultants. I think they hired George a little ahead of the other two of us, and George recommended that they put us on board. So we essentially got in, the three of us, on the ground floor by consulting with that company.... We evaluated this proposal and told the Martin Marietta guys that we didn't think it was up to snuff. We thought that they had some flaws in it that were going to cause it to get rejected. In fact, they told us they didn't really have time to correct those, and they submitted it as it was, and it did get rejected. But then NASA hired a panel of consultants to monitor the progress of the winners. George and Jim and I were all hired to do that.[20]

The winner of the contract competition to design an infrared payload was Hughes Aerospace. The design study was prepared in collaboration with Ames and overseen by the Shuttle Infrared Telescope Science Accommodations Group (SIRTSAG), a review panel of experts formed in 1974 by NASA Headquarters and Ames, which included several people who had participated in earlier panels: David Rank (chair), Eric Becklin, Fred Gillett, D. A. "Al" Harper, Jr. (of Yerkes Observatory), Bill Hoffmann, Frank Low, and Russ Walker (who had left AFCRL and joined Ames in 1976).[21] These scientists had experience across all the infrared techniques that had been tried before—rockets, balloons, aircraft, and the

---

19. According to Witteborn (in a 2 September 2008 interview), the contamination study for which Martin Marietta received $25,000 is L. E. Bareiss, R. O. Rantanen, and E. B. Ress, *Payload/Orbiter Contamination Control Requirement Study Final Report*, MRC 74-93 (Denver, CO: Martin Marietta Aerospace Corp., 1974).
20. Robert D. Gehrz, interview by author, Long Beach, CA, 5 January 2009.
21. Witteborn and Young, "SIRTF" (1976).

ground. They helped set the scientific objectives of SIRTF and lent their expertise on how it might be built. The final Hughes report—"A Large Cooled Infrared Telescope Facility for Spacelab" by Steve McCarthy, Lou Young, and Fred Witteborn—was the first paper to use the acronym "SIRTF."[22] In this original incarnation, SIRTF was designed to be placed on Spacelab aboard the Shuttle. Spacelab was a European project that would provide the Shuttle with a reusable platform for experiments and, perhaps more important, a way for the international community to share the use and costs of the Shuttle.[23] Each Spacelab-equipped Shuttle mission would host a pallet of scientific instruments assembled on the ground and swapped out between missions. Spacelab would use common modules and support systems, thereby increasing its flexibility while reducing costs of the Shuttle as a laboratory. The Hughes study resulted in a design for a cryogenic telescope to be mounted on Spacelab, as well as technical analyses of the thermal and material properties of SIRTF. This study was a key data point in arguments that a Shuttle-borne cryogenically cooled infrared telescope was technologically feasible. But more work was needed to address issues of contamination and viewing time that could compromise its scientific utility.

"After the [Hughes] study, there were more meetings of the infrared astronomers and they became actively involved on a larger scale," Witteborn said.[24]

A key group in this process was the Design Optimization Study Team (DOST), a review panel of experts chaired by Nancy Boggess from NASA Headquarters that included Gehrz, Rieke, and Houck.

This advisory group was expanded into the Focal-plane Instruments and Requirements Science Team (FIRST). Table 4.2 lists the 35 members of FIRST at the time their final report was submitted in 1979.[25] The FIRST group was responsible for developing the language for the Announcement of Opportunity (AO) for SIRTF. The AO is a public document that represents NASA's official commitment to a project and solicits proposals from the scientific and engineering communities to develop that project. "That team concocted the baseline scientific mission and a complement of instruments that it thought were required to accomplish that mission," recalled Bob Gehrz, who was a member of FIRST. Gehrz continued:

> Then it split itself up into subteams that went away and designed the various instruments and wrote conceptual descriptions of them and even defined them to the point where we had optical drawings, tables of filters that would be required, and tables of detectors we wanted to see in them, and sensitivities calculated and everything. That all went into the AO.... They were conceptual designs... meant to be representative of what NASA was thinking it needed for the Shuttle.[26]

---

22. Steve G. McCarthy, Lou S. Young, and Fred C. Witteborn, "A Large Cooled Infrared Telescope Facility for Spacelab," paper presented at the 21st annual meeting of the American Astronautical Society (abstract no. AAS 75–284), held in Denver, CO, 26–28 August 1975; also see Witteborn and Young, "SIRTF" (1976).

23. For more on the history of Spacelab and the Space Shuttle, see *Science in Orbit: The Shuttle & Spacelab Experience, 1981–1986* (Huntsville, AL: NASA NP-119, 1988), available at *http://history.nasa.gov/NP-119/contents.htm* (accessed 30 August 2016); and Roger D. Launius and Aaron K. Gillette, *Toward a History of the Space Shuttle: An Annotated Bibliography*, 2 vols., NASA Monographs in Aerospace History (Washington, DC: NASA History Office, 1992), available at *https://www.nasa.gov/sites/default/files/708235main_Shuttle_Bibliography_1-ebook.pdf* (accessed 30 August 2016).

24. Witteborn interview, 2 September 2008.

25. "Appendices to FIRST Interim Report on SIRTF," Ames Research Center, 14 April 1978.

26. Gehrz interview, 5 January 2009.

TABLE 4.2. Members of FIRST, the committee that developed the initial scientific and technological scope of SIRTF.

Participants in the Focal Plane Instruments and Requirements Science Team (FIRST) and the SIRTF Science Working Group

Name	Organization
Eric E. Becklin	University of Hawaii
Nancy W. Boggess	NASA-Headquarters
Paul E. Boynton	University of Washington
Richard W. Capps	University of Hawaii
K. I. Roland Chan	NASA-ARC
Dale I. Compton	NASA-ARC
Edwin F. Erickson	NASA-ARC
Neal J. Evans	University of Texas at Austin
Giovanni G. Fazio	Harvard College Observatory and Smithsonian Astrophysical Observatory
Robert O. Gehrz	University of Wyoming
Fred C. Gillett	Kitt Peak National Observatory
Doyal A. Harper, Jr.	Yerkes Observatory, University of Chicago
Paul M. Harvey	University of Arizona
William F. Hoffmann	Steward Observatory, University of Arizona
James R. Houck	Cornell University
Roger F. Knacke	State University of New York at Stony Brook
Virgil C. Kunde	NASA-GSFC
Harold P. Larson	University of Arizona
Frank J. Low	University of Arizona
John C. Mather	NASA-GSFC
K. Michael Merrill	University of Minnesota
Alan F. M. Moorwood	ESA
Gerry Neugebauer	California Institute of Technology
Judith L. Pipher	University of Rochester
David M. Rank	University of California at Santa Cruz
Paul L. Richards	University of California, Berkeley
George H. Rieke	University of Arizona
Stephen T. Ridgway	Kitt Peak National Observatory
Baruch T. Soifer	California Institute of Technology
Wayne A. Stein	University of Minnesota
Rodger I. Thompson	University of Arizona
Michael W. Werner	NASA-ARC
Fred C. Witteborn	NASA-ARC
Edward L. Wright	Massachusetts Institute of Technology
Lou S. Young	NASA-ARC

To summarize the events of the 1970s, SIRTF had been formed out of a process that brought together scientists and engineers from universities, government, and industry. Science objectives were developed and prioritized by various panels convened by NASA and the National Academy of Sciences. Industry experts established the feasibility of a cryogenically cooled telescope and produced cost estimates for SIRTF. By the end of the decade, infrared scientists from a range of institutions were in agreement that SIRTF would be worth building.

The next step was to convince Congress to allocate the funds for such a major undertaking. To make the case for SIRTF—as a Shuttle-based telescope—administrators at NASA Headquarters marshaled the evidence from all existing studies on SIRTF's science definition and engineering design. Data supporting this approach were by this time considerable (see Table. 4.3).[27]

---

27. Several of these items are referenced in a presentation on SIRTF's status given by Lou Young, on 24 July 1981 (NASA-ARC, SPS-14, archived in the NASA HRC).

TABLE 4.3. Timeline of key scientific and technical feasibility reports on SIRTF as a Shuttle-attached payload as of 1980.

Date	Activity/Report	Emphasis
July 1969–Oct. 1971	Astronomy and Astrophysics for the 1970s: Astronomy Survey Committee, (Greenstein Report), pub. April 1972. National Academy of Sciences	Policy
July 27–Aug. 15, 1970	Friedman report (1971)—at request of NASA Priorities for Space Research: 1971–1980, Report of a Study on Space Science and Earth Observations Priorities, Space Science Board, National Research Council—recommended that infrared orbiting telescope (70 cm) be studied	Policy
May 1971	Very early system design study: Fred Witteborn, Lou Young. Larry Caroff, Eric Becklin	Engineering
Aug. 1972	Johnson, Rodney O., and Meredith, Leslie. eds. Proceedings of the Space Shuttle Sortie Workshop. Greenbelt, MD: Goddard Space Flight Center, 1974. 2 Vols. (held at Goddard Space Flight Center on 31 July–4 August 1972).	Engineering
Nov. 1972–April 1973	Astronomy Working Group (primarily for Shuttle-borne instruments), chaired by N. Roman, NASA	Science
May 1973	Final Reports of the Space Shuttle Payload Planning Working Groups, Vol. I–V, NASA Goddard Space Flight Center, Greenbelt, Maryland, May 1973.	Engineering
July 1973	National Academy of Sciences. Scientific Uses of the Space Shuttle. Washington, DC: National Academy of Sciences–National Research Council, 1974. (held at Woods Hole in July 1973)	Science
Oct. 1973	Space Shuttle Payloads: Hearing Before the Committee on Aeronautical and Space Sciences. Washington, DC: U.S. Senate Committee on Aeronautical and Space Sciences, 30–31 October 1973.	Policy
Nov. 1973	Naugle, John E. "Research with the Space Shuttle." Physics Today. 26 (November 1973): 30–37.	Science
1974	NASA RFP for facility design study. Martin Marietta hired Gehtz et al to review their proposal draft.	Engineering
1974	$25,000 to Martin Marietta—Bareiss, Rantanen, and Ress, "Payload/Orbiter Contamination Control Requirement Study Final Report" [4]—that was done and finished in 1974.	Engineering
July 1974	Explorer AO, received 13 infrared proposals (Harwit History of IR, p. 40)	Science
Feb. 1974–Jun. 1976	SIRTF Science Accommodations Group	Science
1975	SSB/NAS: IR panel, chaired by G. Neugebauer, of the Space Science Board that met at Snowmass, Colorado, in 1975 (published in Report on Space Science 1975, see below).	Science
Aug. 26–28, 1975	McCarthy, S. G.; Young, L. S.; Witteborn, F. C. (1975) A large cooled infrared telescope facility for Spacelab, Meeting on Space Shuttle Missions of the 80s, Denver, Colorado, Aug. 26–28, 1975, AAS 39 p. (Feasibility and design study by Hughes Aircraft & Grumman Aircraft)	Engineering
1976	"Report on Space Science 1975," National Academy of Sciences, Washington, D.C., 1976.	Policy
1976–1977	Shuttle accommodations and mission system engineering study by Rockwell Intl.	Engineering
Jan. 1976	Witteborn, F. C.; Young, L. S. November 1976 Spacelab infrared telescope facility (SIRTF). J. Spacecr. Rockets, Vol. 13, No. 11 p. 667–674—originally presented at Aerospace Sciences Meeting, 14th, January 26–28, 1976, Washington, DC.	Science

(continued)

Date	Activity/Report	Emphasis
Aug. 1976	"Shuttle Infrared Telescope Facility (SIRTF), Final Report of Preliminary Design Study," Hughes Aircraft Co., Culver City, California, August 1976. Contract NAS2-8494.	Engineering
1977	"Final Report, Vol. 1, Study to Analyze Integration of Shuttle Infrared Telescope Facility (SIRTF)," Rockwell International Corp., Downey, California, 1977.	Engineering
Jan. 1977–Jan. 1979	Focal Plane Instruments & Requirements Science Team	Science
1978	Spacelab Multiuser Instrument Review	Science
July 1978–Dec. 1978	Design Optimization Study Team	Science
April 1978	"Appendices to FIRST Interim Report on SIRTF," Ames Research Center, 14 April 1978.	Science
1978–1979	Telescope design optimization study by Perkin-Elmer, Beech Aircraft & SAI	Engineering
Jan. 1979–Nov. 1979	SIRTF Science Working Group	Science
1979	CSAA, Space Science Board, NAS	Science
Aug. 1979	SIRTF System Design Summary Document, Final Report of Design Optimization Study, Perkin-Elmer Corp., Norwalk, Connecticut, August, 1979.	Engineering
Sept. 1979	"SIRTF Design Optimization Study: Final Technical Report," Perkin-Elmer Corp., Norwalk, Connecticut, 30 September 1979. Contract NAS2-10066.	Engineering
Nov. 1979	Final Report of the Focal Plane Instruments and Requirements Science Team and the SIRTF Science Working Group, NASA Ames Research Center, November, 1979.	Science

Attached to the Shuttle

To persuade Congress to authorize funding for SIRTF, scientists had to demonstrate that the project would meet a strong need. But first, the scientists themselves had to be convinced that SIRTF was worth doing, meaning that it was the best use of the nation's shrinking research funds. Scientists and Congress wanted to know: Should we spend our money on SIRTF?

The decadal survey published in 1972 (the Greenstein report) had shown how effective a consensus document could be for the field of astronomy. Although there was much internal debate, through the Greenstein report the scientists showed a largely united front that made it clear which projects NASA and Congress should fund. On the basis of the National Academy of Sciences' recommendations, scientists saw funding flow in the 1970s toward major missions, such as the Large Space Telescope (Hubble) and the Compton Gamma Ray Observatory (CGRO), and toward moderate (under $200 million) Explorer missions, such as the Infrared Astronomical Satellite (IRAS), an international collaboration to conduct an infrared sky survey. However, the 1970s was an economically difficult decade for both NASA and the country, and not every project the scientists recommended was funded. While several ground-based observatories were built (for optical, radio, and infrared wavelengths), the budget for airborne experiments shrank. With limited funding for rockets, balloons, and aircraft, infrared astronomy was still operating mostly from the ground, through a layer of atmospheric dust and water vapor. The only politically viable way to get off the ground was to be on the Space Shuttle, even if it, too, had a layer of dust and water vapor.

In this environment, the 1980s started with both a new decadal survey by the National Academy of Sciences and a commitment by Ames to form an official SIRTF study team tasked with producing a preliminary design based on the work of the FIRST group.[28] But just as things were starting to come together, the project began moving in two directions. The NAS decadal survey committee took the position that SIRTF should be a Shuttle-attached telescope—as all of the studies in the prior decade had assumed it would be. However, the scientists who had done those studies were increasingly advocating for SIRTF to be a free flyer: the Shuttle might be used to put SIRTF in orbit, but in their view SIRTF should be left in space when the Shuttle returned to Earth. As a stand-alone satellite, SIRTF would be more scientifically useful but far more costly. Only one of these two options could be supported. Resolution of this scientific and economic tension would come while NASA was preparing the Announcement of Opportunity.

The Announcement of Opportunity for SIRTF

Ames created the official SIRTF study group in January 1980, with Fred Witteborn as the study's lead scientist and Lou Young as its manager. Among the group's goals were "to force the release of the Announcement of Opportunity, and … to make sure that funding continued for more studies," Witteborn said, adding:

> I think the driving force behind this [AO] was the infrared astronomers, the infrared community. It was organized by Ames at that time, but also [NASA] Headquarters. Nancy Boggess and Nancy Roman were the Headquarters persons involved.… This group really was mainly focused on defining the scientific requirements better and giving the rationale for why you needed the infrared telescope. They did a lot of calculations, and there was some real science behind it. They were trying to show why you needed [an infrared telescope] in addition to the Hubble telescope. Congress's point of view was, you've got a bunch of telescopes already, what more do you want?
>
> We had a series of meetings and the whole thrust of all of these meetings and studies was to get the Announcement of Opportunity out. By 1980, infrared astronomers were quite sure that they wanted this telescope. The next step was to get NASA to get committed to funding. The way you do that is you get NASA to release an Announcement of Opportunity. Before NASA would do that, they wanted to be very sure of what it was going to cost and what the science was going to be like. They had to compare it with other projects. By then, there were all sorts of scientists vying for funding, particularly astronomers—UV astronomy, gamma-ray astronomy, x-ray astronomy, solar astronomy—all these people wanted major projects, minor projects. And from their points of view, they each had a better reason for doing their projects than the others. So the period from 1980 to 1984 was really spent in refining the requirements and explaining the possibilities as to what the science accomplishments could be, what the science objectives were, as well as the mission requirements.[29]

---

28. Director's memo AO-15, 31 January 1980, NASA Ames History Office, NASA Ames Research Center, Moffett Field, CA, PP05.04, Larry A. Manning Papers 1967–1988, Box 2, Folder 3.

29. Witteborn interview, 2 September 2008.

Meanwhile, the National Academy of Sciences convened the Astronomy Survey Committee, under the chairmanship of Harvard astronomer George Field, for another decadal survey to establish the science priorities for the 1980s.[30] As a member of the decadal survey committee, Bob Gehrz was told to assume that SIRTF was safe to move forward as a Shuttle-based project and would not need prioritization. Although Gehrz knew that within the SIRTF working groups they were moving away from this configuration, he could not take this factor into account. Gehrz explains:

> It was initially planned that this [Shuttle-borne] telescope was going to launch for the first time in about 1979 and that the price tag would be between $100 million and $120 million. It would be a 120-centimeter telescope. When it came time to do the decadal astronomy survey chaired by George Field for the decade of the '80s, NASA told that committee that SIRTF should be considered a project that was safe from prioritization.... So it didn't actually get a prioritization with all the projects in that decadal survey that were to compete against one another; it was one of the baseline projects that NASA already had in hand. At that same time, some of the people on the SIRTF project—led by Frank Low, as I recall—began to agitate for it being put into Earth orbit [as a free flyer], because it was recognized that the Shuttle environment was much too dirty to really fly a cryogenic telescope. It was also recognized at that point that bringing a telescope like that down through the atmosphere was going to contaminate it so badly that it would essentially have to be rebuilt from scratch every time you reflew it. It would have to be completely disassembled and cleaned, it would be so badly contaminated. So they began to agitate for it being a telescope that was launched from the Shuttle bay but then remained in Earth orbit, where it could be refurbished periodically by the Shuttle. Then the price began to go up a lot, and it also changed the baseline concept to the point where NASA said, "This is no longer the done deal that was presented to the Field committee. This has to be reprioritized." So it threw the project into a whole new regime....
>
> I was on the Field committee, and I caught a lot of flak about this, because some of the people on the project claimed I hadn't advocated it strongly enough to get it into the review process. But we had been told we couldn't put it in there. Frank Martin was the guy at NASA who pretty much steadfastly held that line. So it wasn't prioritized in the Field committee as a Shuttle-attached mission, and it certainly wasn't in there as a free flyer. So that meant it was a project that was now off the charts. It was a brand new project, which had to be reprioritized by the next decadal survey [for the 1990s].[31]

The project NASA was willing to fund was a Shuttle-based SIRTF. The project the scientists wanted was a free flyer. Both sides were guilty of wishful thinking. The net result, however, was that the Field report did not explicitly prioritize

---

30. This survey, which would become known as the Field report, was published under the title *Astronomy and Astrophysics for the 1980's, Volume 1: Report of the Astronomy Survey Committee and Volume 2: Reports of the Panels* (Washington, DC: National Academies Press, 1982–1983). Volumes 1 and 2 are available online at *http://www.nap.edu/catalog.php?record_id=549* and *http://www.nap.edu/catalog.php?record_id=550*, respectively.

31. Gehrz interview, 5 January 2009.

SIRTF in the decadal survey and so offered no political cover for the project when SIRTF later changed scope.

A Spectrum of Projects

Frank Martin, who had told Gehrz that SIRTF was a done deal, ultimately decided the priority of NASA's astronomical missions. He had been playing that role since the 1970s, when he was put in charge of advanced programs in the Astrophysics Division and had overseen the selection of some smaller-scale infrared projects that included the Cosmic Background Explorer (COBE) and the Infrared Astronomical Satellite (IRAS). COBE would test the Big Bang theory by measuring the cosmic microwave background. IRAS was developed and operated by astronomers from the United States, the Netherlands, and the United Kingdom to conduct an infrared sky survey. Less complex than SIRTF, COBE (launched in 1989) and IRAS (launched in 1983) had small budget profiles and were being built as free flyers. Martin was committed to seeing both of these projects through.[32]

Around 1979, Martin was promoted to Director of the Astrophysics Division. Like everyone appointed to that position, he received a never-ending stream of visitors who argued that their pet projects should receive priority. Scientists typically felt that more money should be spent for research in their favorite wavelength—optical, x-ray, gamma ray, ultraviolet, radio, and increasingly, infrared. It had not taken long for NASA to adopt a portfolio of projects to support across the electromagnetic spectrum. Funding and personnel were limited, however, as both the Carter and Reagan administrations cut space science budgets. Martin was forced to cancel projects in order to keep the budget in line, terminating missions already in orbit (Copernicus and Einstein) in favor of those being jointly developed with the European Space Agency (International Solar Polar Mission and a Halley's Comet flyby). This only increased the number of scientists who lined up at his door to either request new projects or reinstate canceled ones.

In addition to balancing the needs of the scientific community, Martin was finding it increasingly difficult to explain to Congress why NASA needed all these projects. It was hard to sell politicians on these projects one at a time, especially when the projects were neither distinct enough from one another nor similar enough to speak of them as a package. It was a hodgepodge of projects—all possessing scientific merit and passionate advocates—and Martin had to find a way to ensure that the right projects got the right resources at the right time.

IRAS and COBE were high on Martin's to-do list, and the National Academy of Sciences agreed that these projects should receive priority. However, Martin's peers in the Defense Department were unconvinced that this was money well spent. Nancy Boggess remembers the day she represented NASA at a meeting at the Pentagon, before IRAS launched:

> I gave a presentation on what IRAS was to do. I gave a big long spiel, and at the end of it this admiral stood up and said, "Do you mean to tell me you put a bucket of superfluid helium up there with a telescope in it and expect to see anything but white light?" I said yes. The entire table guffawed; they agreed with him. That didn't make NASA feel very comfortable—that all these big military men were thinking we wouldn't have a prayer of a chance of seeing anything with IRAS. How wrong they were![33]

---

32. Frank Martin, telephone interview by author, 27 March 2009.

33. Boggess interview, 19 March 2009.

Launched in January 1983, IRAS produced an extremely successful sky survey, a project that the Air Force had unsatisfactorily pursued for two decades. But until the results were in, this skepticism did not make Boggess's or Martin's jobs any easier.

Something that did help Martin was an article by Martin Harwit.[34] Published in 1975 (and expanded into a book in 1981), Harwit's thesis argued that technology was the driving force in astronomy. "Somebody walked down the hall and handed it to me," Frank Martin remembers:

> I read it and I knew exactly what to do once I read that article. I told Harwit he wrote this for two people and probably only one—it was me—it was the Director of Astrophysics at NASA. Because many of these things that he talked about couldn't be done from the ground. He made it very clear about the criteria for determining what kinds of mission should be pursued. Every part of the electromagnetic spectrum. It's a very clear discussion.[35]

Ironically, Harwit had written the book after NASA cut the funding for his sounding-rocket program, which increased his free time.[36] Frank Martin describes how Harwit's ideas shaped his thinking:

> I had a mindset that these things were all expensive. Whether they were the small Explorer missions, or Hubble Space Telescope, or something that was supposedly going to fly on the Shuttle, like SIRTF—these things were really expensive. Yet the science of these missions was necessary.... But what I thought, as a representative of the taxpayers' money, was that there had to be a strong discovery element in all these missions, given the fact that we really hadn't looked at the universe across the whole electromagnetic spectrum. Harwit's article really spoke to me—that's what we should be doing. So whether we were looking at... attached payloads in the Shuttle, which SIRTF was in the early phases, or at Hubble Space Telescope, the gamma-ray observatory, the Chandra X-ray Telescope, all those things were concepts in the mid-70s that we funded and defined when I was doing advanced programs. All those things were pieces of this huge puzzle that... Harwit talked about in his book.[37]

Four telescopes representing the essential wavelengths and providing orders-of-magnitude greater resolution than any other instrument before them were already under way at NASA. Harwit had provided a compelling story for cosmic discovery. "We were able to package all this stuff together," Frank Martin recalls, "in such a way that folks in Washington, DC, and at NASA could understand that a telescope wasn't a telescope wasn't a telescope—what we were doing was something that could only be done by NASA, and could only be done by the federal government, which was to open up this huge discovery space for science and the country."[38]

---

34. Martin Harwit, "The Number of Class A Phenomena Characterizing the Universe," *Royal Astronomical Society Quarterly Journal* 16 (December 1975): 378–409. See also Martin Harwit, *Cosmic Discovery: The Search, Scope and Heritage of Astronomy* (New York, NY: Basic Books, 1981).
35. Martin telephone interview, 27 March 2009.
36. Harwit interview, 26 May 2009.
37. Martin telephone interview, 27 March 2009.
38. Ibid.

As for SIRTF, the infrared was going to be represented first by IRAS and COBE. "I think people were looking for payloads to fly on the Shuttle, and in my mind we were busy focusing on getting IRAS flown," Martin says:

> That was one of my passions, that and COBE—during this period as Director of Astrophysics—getting that thing built.... Between spending Explorer money on IRAS and getting the COBE mission going, I knew that I wasn't going to get the money to do SIRTF for some time, and all I really felt was important to SIRTF was that we kept it going and kept it alive. Whether it actually flew on the Shuttle wasn't that important to me at that point. What was important was that I had a source of funds that allowed us to continue to work on the technology and continue the planning.... The nice thing about the Shuttle-attached payloads was that it gave me another source of funds to do that. The agency [NASA] was willing to put money into those kinds of things. So during the late '70s and very early '80s, I was content to let the science team and Ames continue to work on SIRTF as an attached payload, and let's worry about what we're really going to do when we get to the point where we're really talking about moving out with it.[39]

It all came together at the start of 1983. Congress had recently granted funding approval for COBE, and IRAS launched on 25 January. Shortly thereafter, with his goals accomplished, Frank Martin left NASA Headquarters. Charles Pellerin, who had been Martin's deputy, took over the role of Director of Astrophysics. Support at Headquarters for Shuttle-based projects was extremely high. However, Pellerin felt that it was a poor use of funds to attach SIRTF to the Shuttle, where it could gather only a few days of data at a time. Nancy Boggess and the small community of infrared scientists agreed. Nevertheless, Boggess persuaded Pellerin that they should proceed with what was sellable to NASA administrators at that moment and change SIRTF to a free flyer later. On 13 May 1983, Pellerin authorized the Announcement of Opportunity for SIRTF as a Shuttle-attached facility.[40]

The AO called for proposals from the scientific and engineering communities to design and build instruments for the first-ever space-based infrared telescope facility. The AO stated:

> SIRTF is envisaged as an attached Shuttle mission with an evolving scientific payload. Several flights are anticipated with a probable transition into a more extended mode of operation, possibly in association with a future space platform or space station. The SIRTF will be a 1-meter class, cryogenically cooled, multi-user facility consisting of a telescope and associated focal plane instruments. It will be launched on the Space Shuttle and will remain attached to the Shuttle as a Spacelab payload during astronomical observations, after which it will be returned to Earth for refurbishment prior to re-flight.[41]

It had taken 12 years to get to this point—the AO reflected the alignment of political, economic, technical, and scientific arguments. It was a fragile conjunction. For a moment in May 1983, Fred Witteborn could be glad that NASA had released an AO for SIRTF, a public

---

39. Ibid.
40. Charles J. Pellerin, interview by author, Boulder, CO, 19 March 2009.
41. NASA Announcement of Opportunity OSSA-1-83: Shuttle Infrared Telescope Facility (SIRTF), 13 May 1983.

declaration that it was supporting this project. SIRTF had even won over the decadal survey committee, which had withheld support in the 1970s and now granted SIRTF full support (as a Shuttle-attached payload) in the 1980s.

The decadal survey for the 1980s, NASA's budget for infrared science, and the AO for SIRTF were obsolete by July. As early results from IRAS came in, it became clear that an infrared telescope had to be a free flyer to fully realize the potential scientific benefit. Frank Low, in particular, took to this argument and lobbied NASA to change the AO. By 22 November, the last drop of cryogen was used by IRAS, and its mission was completed. Low's mission was also achieved. The AO had been amended, and the proposals arriving at NASA by the 5 December deadline included instrument designs for SIRTF as a Shuttle-based telescope with options as a free flyer.

CHAPTER 5

Selling It

The Announcement of Opportunity provided the SIRTF project with the legitimacy and visibility to recruit industrial and academic partners to design and build it. Attracting those partners is one thing; managing a growing project is another. As a NASA project, SIRTF acquired resources through a key management process colloquially called "selling it," which involved simultaneously achieving priority in the scientific community, at NASA Headquarters, and in the national budget.

To manage large projects, NASA uses a system that is both hierarchical and decentralized. Hierarchically, Headquarters exercises oversight at the program and project levels and selects a program scientist from Headquarters' staff to oversee one or more projects. At the next level down, a project scientist and project manager are selected from the staff at one of the NASA Centers. The project scientist is responsible for working with the Headquarters program scientist to develop the core scientific objectives of the mission (the "Level 1 Requirements"), while the project manager ensures that those core objectives are met and that costs and schedules stay in line during design and construction.

Recognizing that hierarchical controls are insufficient in projects of great complexity, Headquarters turns to experts—from industry, academia, and other NASA Centers—to compose the project teams. NASA retains oversight through its program and project leaders. Nancy Boggess, SIRTF's program scientist (a role she also held on other infrared projects, including IRAS, IRS, and COBE), explains the job thusly: "I represented NASA, and I was to get an agreement from the scientific community—that they would stand by this project as it evolved and come up with good ideas to reduce the cost and do things that would make it work well."[1] The SIRTF project team members were selected on the merits of their instrument and science proposals, which were prepared as formal responses to the Announcement of Opportunity released by NASA Headquarters in May of 1983.

Selecting the Team

The AO for SIRTF focused on the science that such a telescope might achieve and called for two types of proposals. The first was for "focal-plane investigations." The focal plane is the area of the

---

1. Nancy Boggess, telephone interview by author, 2 March 2009.

telescope where infrared flux is collected and recorded by instruments tuned to detect photons or spectra. The second was for investigators to oversee the project, either as "facility scientists," responsible for ensuring that the overall design of the telescope facility would address the project's core scientific questions, or as "interdisciplinary scientists," theoreticians or generalists who would maximize the applicability of SIRTF's data to users across the astrophysical disciplines.[2]

Instrument proposals were submitted by several teams. Eric Becklin led a team from the University of Hawai'i to build a near-infrared camera, with Honeywell as the prime contractor for the instrument and Rockwell building the detectors. Bob Gehrz, in collaboration with Larry Caroff and Ball Aerospace, also proposed a near-infrared camera. Fred Witteborn participated in a proposal for a spectrometer to measure the chemical composition of stellar targets.[3]

Despite the active role each of these scientists had played in early design studies, neither Becklin, Caroff, Gehrz, nor Witteborn had their proposals selected for funding. Gehrz recalls,

> We went to NASA for our debriefing. We were told by Nancy Boggess and her colleague Charlie Pellerin. They told us that we were the No. 1 science team and that we had simply been deselected because we had no experience building space hardware, which I found to be amusing when it became my turn to actually be facility scientist for this telescope. I really did think I knew how to do that kind of stuff.... The point is, I had never done it. NASA's argument was that I shouldn't be allowed to do it, therefore. Only people who had done it should be allowed to join the club.[4]

The list of infrared scientists was short in 1984, and the list of those with space-hardware experience was even shorter.

The winning proposals had space experience or were in the process of getting it. Giovanni Fazio, of the Harvard-Smithsonian Center for Astrophysics, was named Principal Investigator (PI) for SIRTF's Infrared Array Camera (IRAC). Fazio was concurrently the PI for the first Shuttle-based infrared instrument (IRT), which would launch on the Spacelab-2 pallet in July 1985. Jim Houck, of Cornell, was named PI for SIRTF's infrared spectrograph (IRS). Houck was a member of the science team for IRAS, the international sky-survey project that had launched in January 1983. And George Rieke, of the University of Arizona, was named PI for SIRTF's far-infrared camera, or Multiband Imaging Photometer (MIPS). Rieke didn't have space experience, but he submitted what may have been the only proposal for a far-infrared camera, and he worked closely at the University of Arizona with Frank Low, who was part of the IRAS science team.[5]

In addition to the instrument selection, NASA named Frank Low as SIRTF's facility scientist,

---

2. NASA Announcement of Opportunity OSSA-1-83: Shuttle Infrared Telescope Facility (SIRTF), 13 May 1983.

3. NASA does not publicly disclose the names of all those who submit proposals, only of those whose proposals are selected. Thus, this is a partial list, as it includes only the names of those who indicated that they responded to the AO. The list is based on the following sources: Eric Becklin, interview by author, Long Beach, CA, 5 January 2009; Robert D. Gehrz, telephone interview by author, 20 January 2009; and Witteborn interview, 2 September 2008.

4. Gehrz telephone interview, 20 January 2009.

5. NASA Ames Press Release 84-32. An additional reference is "A Short and Personal History of the Spitzer Space Telescope," by Michael Werner, 1995, ASP Conference Series, preprint available at *http://arxiv.org/PS_cache/astro-ph/pdf/0503/0503624v1.pdf* (accessed 30 August 2016). The U.S. IRAS science team consisted of G. Neugebauer,

and Mike Jura and Ned Wright, who had both received their doctorates from Harvard and were now professors at UCLA, as multidisciplinary scientists. Jura and Wright, between them, represented a wide swath of astrophysical interests, as Jura's work focused on processes within the solar system, and Wright's focused beyond it. Ames was designated the host Center; from Ames, Michael Werner was selected as project scientist, Fred Witteborn as deputy project scientist, and Lou Young as project manager.

Managing Influence

It may seem surprising that Fred Witteborn was not made SIRTF's project scientist—after all, he had led the project up to that point, and people liked him. But there were two factors that diminished his influence: He was a member of the infrared community, and he worked at Ames. Both were battlefields.

The infrared community was more contentious than other astronomy fields. Perhaps this was because infrared astronomy was so new that little agreement existed on what was the right thing to do or the right way to do it. Without a paradigm, scientists may not readily agree on a direction. When seeking the consensus needed to move forward, forceful personalities were often more persuasive than facts. As Witteborn recounts:

> I was at a meeting at Snowmass [in Aspen, Colorado], and we were trying to develop priorities. One of the Headquarters people who was from the Solar Astronomy Group was also having solar astronomy people meet at Snowmass at the same time. The Solar Astronomy manager came over to our meeting, which was being led by Nancy Roman and Nancy Boggess. He came over to our meeting and listened in. Listened to us fighting. He finally said, "Let me tell you how our group works. They get the funding every year." If you look at the history of NASA [in solar astronomy], there is OSO-1, OSO-2,... [OSO-3 and up through the Explorer series]—when I started looking at the numbers, they were up to OSO-35.... I mean, they had lots and lots of projects. They ordered them, and they got together, and they had friendly meetings. He said, "Contrast that with this group. You're fighting each other tooth and nail. Each one of you is saying that more should be done at your institution and not at the other institution." I may be giving the false impression that they were fighting with us. The [infrared scientists] were fighting with each other.... One of the Headquarters directors in the early days would go around to each of the places where Headquarters' astronomy money was being spent. She would talk to the people there about their needs for the next year and so forth. Headquarters' science managers would travel from university to university, from Center to Center. Then, of course, budget restrictions got such that Headquarters people had to save their travel and they couldn't travel anymore. But anyway, she came here [to Ames] after she went to the University of Arizona. She said, "Well I fund six groups there in astronomy and infrared astronomy. So I asked them to all meet with me together. But they're not talking to each other, so we had to meet

---

Chairman, H. H. Aumann, C. A. Beichman, T. J. Chester, T. N. Gautier, F. C. Gillett, M. G. Hauser, G. Helou, J. R. Houck, C. J. Lonsdale, F. J. Low, B. T. Soifer, R. G. Walker, and E. T. Young. This list appears in the acknowledgments section of the *Explanatory Supplement to the IRAS Sky Survey Atlas*, by Sherry L. Wheelock et al. (Pasadena, CA: Jet Propulsion Laboratory, 1994), available at *http://irsa.ipac.caltech.edu/IRASdocs/exp.sup/ch13/A.html* (accessed 30 August 2016).

separately." So you could see how tight the competition was.[6]

In such a contentious environment, prestige determined the degree to which one's voice was heard. Those with the most prestige came from universities; those with the least came from government. And within NASA, the Centers that managed space missions (Goddard, Marshall, Johnson, and JPL) had more institutional clout than Centers that focused on aeronautics (Ames and Langley).

This pecking order was especially true in the 1980s, when Ames found itself at the bottom and competing with the other Centers for resources and respect. Although it is one of the oldest NASA Centers, Ames is one of the most vulnerable to political and economic forces. At its inception during World War II, Ames had been intended to serve as a backup facility for Langley Research Center, in Hampton, Virginia, in case the East Coast was attacked. After the war, Ames was often viewed by Headquarters as an outpost that could be closed down. "From the Eastern point of view, Ames was always a place that was a redundant system that they set up because of the international situation," Witteborn says. "As Ames continued to grow, it was more and more competitive with the East Coast, or at least began taking away their resources. So it was obviously a target for closure." Ames's relative redundancy was compounded by its isolationism. The early history of Ames depicts an organization that avoided interactions with Headquarters.[7]

Without strong ties to or clear support from Headquarters, Ames directors were either disinclined or unable to build the teams and capabilities for complex projects. "The Pioneer program [a series of spacecraft designed to explore the planets, which Ames managed and which ran from 1958 until 2003, when Pioneer 10 sent back its last data] was done with very little enthusiasm from the outside, even [from] the Ames directors," Witteborn recalls.[8] Administrators at Ames shied away from large projects.

As a result, Ames lacked not only prestige but also resources. Throughout the 1960s, when the rest of NASA was expanding rapidly to support the Apollo Moon missions, Ames did not share in this growth. Ames focused on aeronautical research, and its role in Apollo was mostly confined to working on reentry-vehicle dynamics and nose-cone design. Then, as the Apollo program was nearing its goals, NASA started to scale back on the research effort. The first hiring freezes were in 1965. "It comes back to haunt you as time goes on," Witteborn says. He continues:

> When the Astrophysics Branch was set up here [in 1971], why didn't we hire any astronomers? It wasn't because we didn't like astronomers. It was a hiring freeze. It was very difficult. One of the most useful people I had was another physicist who, like myself, was very good at [designing] instruments. And of course, he was supposed to be designing instruments for airborne studies. His name was Ed Erickson. He was a

---

6. Witteborn interview, 2 September 2008. Witteborn referred to Explorer 35 as "OSO-35," but in fact solar astronomers had two programs—the Orbiting Solar Observatories program, which went up to OSO-9, and the Explorer program, which went up to Explorer 59. Simpler to build than an OSO, the Explorers were comparatively inexpensive and were used to conduct a range of experiments that were largely dominated by solar science. For more details, see *http://history.nasa.gov/explorer.html* and *http://www.jpl.nasa.gov/missions* (accessed 30 August 2016).

7. This characterization of Ames is supported in interviews conducted by the author and is detailed in chapter 2 of Glenn E. Bugos, *Atmosphere of Freedom: Sixty Years at the NASA Ames Research Center*, SP-4314 (Washington, DC: NASA, 2000), available at *http://history.nasa.gov/SP-4314/sp4314.htm* (accessed 30 August 2016).

8. Witteborn interview, 2 September 2008.

postdoc, and when his postdoc ended, [as branch chief] I wanted to hire him. I was told there were no positions. I said, "Come on now, this is really important." They said, "Well, we know we need to hire some people at Ames. What Ames managers had done was to issue hiring warrants to branch chiefs, and if you get one, then you can use it any way you want. If you want another one, they are going to ask you how the first one went." So it was really tough to hire Ed. I immediately announced my intentions that I wanted to hire him. Then I left for a meeting at Headquarters for a few days of planning for infrared astronomy related to the LST [Large Space Telescope, which became the Hubble] project. When I came back three days later, they said, "A funny thing happened while you were gone. We lost your hiring warrant. We hired somebody else."

Eventually, after several more warrants and an intervention by Hans Mark, the Director of Ames at that time, Witteborn was able to hire Erickson.[9]

Acquiring resources for a major new program like SIRTF in an environment of scarcity is a process fraught with internal competition and political maneuvering. Solar astronomy was less contentious because they had history on their side—the Sun was a well-established target for observations, most obviously in the traditional, visible wavelengths. Infrared was a new field and lacked legitimacy.

Witteborn mastered this game at least partially, as by 1980 there were 14 infrared astrophysicists working at Ames.[10] But this was of little import to the infrared community. What mattered was who those astrophysicists were. According to Witteborn, "The way the university science works in infrared astronomy and in cryogenics as well [is] if you have some of their doctoral students here, why, then you are recognized." He continues:

> The astronomical infrared community did not see us as major players. In my branch, my first question when I was asked to be branch chief, which was in 1971, was, How many astronomers can I hire? Mike Bader and Hans Mark said, "What? You got two in the branch." But these weren't major astronomers, they were grad students really.... [I]t was a long time before we actually were able to hire an astronomer outright. However, we did have the Theoretical Branch..., which had a number of astronomers. Larry Caroff was one of them. And they had Jim [James B.] Pollock, who suggested projects to study from the Learjet. In fact, he and I and our collaborators here found sulfuric acid in the Venus atmosphere. Now, three groups simultaneously discovered that—two working from the ground and one from the air—but we had the most complete spectrum, because we could get above the water vapor in Earth's atmosphere. And that led to early discoveries.

> But anyway, we were not looked upon by the infrared community as good astronomers. They wouldn't tell *us* that; they told Headquarters. And Headquarters said, "You guys aren't producing any papers." But that wasn't entirely true. At one meeting, the session chairman commented on our paper on far-infrared interferometer spectra of the Orion Nebula and said, "Well, the shape is not particularly exciting, but this is the first far-infrared spectrum of a nebula that was ever taken. It's interesting from that point of

---

9. Witteborn interview, 2 September 2008.

10. SIRTF System Summary Review, Ames Research Center, 24 July 1981, slide 13.

view. It's a continuum." I won't tell you how hard it was—how hard we had to work before we produced that big blob [spectral graph], but there it was. So then you hear[d], "Well, you guys aren't using the big telescopes." We had already developed the instruments, so we applied for [observing] time. We got time on big telescopes. So the next thing was "Well, those aren't big enough telescopes." So we got time on bigger telescopes. We got time on the IRTF [InfraRed Telescope Facility at the University of Hawai'i], which was the big infrared telescope at that time.[11]

Yet it was never enough. The infrared community, dominated by university scientists, measured success by the criteria used in universities, all but guaranteeing that Ames scientists would fall short. It did not matter that Witteborn had a doctorate from Stanford. He had spent his entire career at Ames, and his research was constrained by its institutional realities—working on missions that aligned with NASA's goals and that were within the budgets that Congress authorized. And he was further constrained by the low status of Ames within NASA, the historical result of Center management that positioned Ames for aeronautics and not NASA's high-profile plans for space-based projects.[12]

Selecting a Leader for SIRTF

Despite Ames being a relatively weak platform from which to sell a new project, the scientific community saw SIRTF as the major infrared mission. This was the message delivered in the National Academy of Sciences' decadal survey for the 1980s, known as the Field report and authored by eminent (and university-based) astronomers and astrophysicists.[13] Naturally, it would matter to them who would be selected as the SIRTF project scientist, for this person would represent the scientific concerns of the community and ensure that the data obtained by SIRTF would be of high quality. "In the early '80s," Witteborn reports, "I heard indirectly that the astronomical community told Headquarters—who then talked to upper management [at Ames]—that they'd really like the job that I was doing to be done by 'somebody from the infrared community.' So the job opening was created and they invited people from everywhere.... [M]y bosses offered it to me, I think, as a formality. I told them that the [infrared] community would be more comfortable with Mike Werner, and he became SIRTF project scientist."[14]

Mike Werner had joined Ames in 1979, but since earning his doctorate under Martin Harwit at Cornell, he had worked in some of the most prestigious astronomy circles. The route he took to get there included some familiar paths. Originally starting out as a physics major, Werner switched to astronomy while an undergraduate at Haverford College, where Nancy Boggess had taken master's-level courses in astronomy a few years earlier.[15] To gain more experience in astronomy before entering graduate school, Werner spent a year at the Naval Research Laboratory.

---

11. Witteborn interview, 2 September 2008.

12. The situation that Ames faced in the 1960s, 1970s, and 1980s is somewhat changed today. Ames has remained small by comparison with other NASA Centers (which tend to be better positioned for managing large-scale programs); however, Ames has emerged as a valuable research center and contributor to innovation in Silicon Valley and the aerospace industry.

13. Field report, passim.

14. Witteborn interview, 2 September 2008.

15. John C. Mather and John Boslough, *The Very First Light: The True Inside Story of the Scientific Journey Back to the Dawn of the Universe, rev. ed.* (New York: Basic Books, 2008), p. 90.

There he met Harwit, who was at the NRL on a fellowship and encouraged Werner to apply to the Cornell doctoral program. Harwit became Werner's thesis advisor, and after graduating from Cornell in 1968, Werner spent a year in Cambridge, England, at Fred Hoyle's Institute for Theoretical Astronomy. Hoyle, a well-known theorist, had hosted Harwit as a postdoc in the early 1960s.

Returning to the United States, Werner accepted a postdoc appointment at Berkeley (1969–1972) in the lab of Charles Townes, who had received the 1964 Nobel Prize in physics for his work on masers and was now developing their application to questions in infrared astronomy.[16] Townes describes one of the early projects he and Werner worked on, showing how necessary it was to build one's own equipment and how skepticism had to be applied to the results:

> We were fortunate to begin almost immediately with what seemed like an interesting and relatively simple measurement, even though a new spectrometer needed to be put together for it. Martin Harwit and his associates at Cornell had just published observations from high-altitude rocket observations indicating a remarkably intense isotropic flux in the 0.4–1.3-mm wavelength range.... The flux was about 25 times more than expected from a 2.7-K blackbody field. It appeared strong enough and at wavelengths where the atmosphere was transparent enough that we should be able to observe the radiation from a high-altitude location on Earth. Mike Werner, a newly arrived postdoc at UC Berkeley, and John Mather, a graduate student interested in our research, were willing to try to measure this exciting but very puzzling radiation. Could it even be some intense spectral line? I asked Paul Richards, a fellow professor experienced with bolometers and far IR, for help to speed up the work, and fortunately he was also interested. These three put together both a tunable and a fixed Fabry-Perot interferometer with nickel mesh reflectors, an indium-antimonide bolometer detector, a chopper, and a focuser of 8-cm aperture.... The system was set up at a 12,500-foot altitude on White Mountain in eastern California, and spectra were taken in the 0.7–1.7-mm wavelength range with a resolving power of about 100. The radiation apparently detected by rocket flights didn't seem to be there! Rocket measurements are of course difficult, and this was not the only time that rocket measurements were to give misleading results in measuring the isotropic background radiation.[17]

Werner's two stints as a postdoc, at Cambridge and Berkeley, gave him an opportunity to work with prominent researchers in both theoretical and experimental astrophysics. While still at Berkeley, he began conducting experiments on the Learjet at Ames (and later, on Ames's Kuiper Airborne Observatory). This airborne work continued when he joined the Caltech faculty

---

16. Michael W. Werner, interview by Sara Lippincott, Pasadena, CA, 25 July 2008, Oral History Project, California Institute of Technology Archives; a 41-page transcript of this interview is available online at *http://oralhistories.library.caltech.edu/163/* (accessed 30 August 2016).

17. Charles H. Townes, "A Physicist Courts Astronomy," prefatory chapter in the *Annual Review of Astronomy and Astrophysics* 35 (1997): xiii–xliv. Townes continues: "The work clearly interested Paul Richards, who then moved into rocket measurements of the background radiation. John Mather became Paul's student, and eventually was to lead a spectacularly successful experiment, with the COBE satellite, to measure the background radiation and apparently really get it right" (p. xxxiv). Mather did indeed get it right—he won the Nobel Prize in physics in 2006 for his work on COBE, for which he served as the project scientist.

in 1972.[18] As part of the infrared group with Gerry Neugebauer and Eric Becklin, Werner continued to develop instrumentation and pursue research that included observations of Orion in the mid-infrared.[19]

SIRTF, of course, was just getting started. Becklin, with the encouragement of Neugebauer, had gotten involved in SIRTF, which is how Werner had learned of the project. "[Becklin] was on one of the very early study teams," Werner recalls. "I remember him talking about that at one of our sack lunches up in 469 Lauritsen [at Caltech], which we had every week. But in January 1977, because of the success of my work on the Kuiper Airborne Observatory, I was asked [by David Rank, at UC Santa Cruz, whom Werner knew from Berkeley] whether I wanted to head a sub-team of one of these study groups that was helping to define SIRTF. So I said, sure, I'd like to do that."[20]

The timing was fortuitous. Werner received an unfavorable tenure decision from Caltech, so he joined Ames in 1979 at the invitation of Fred Witteborn. Thus, when the time came to select a project scientist from among the NASA ranks, Werner had both the requisite knowledge about the project and academic experience at premier institutions. He also knew how to build instruments, a skill born of necessity from working in infrared astronomy. Finally, Werner had an almost egoless quality that would serve him well in the often-contentious infrared community—he made people feel connected to the project without putting himself at the center. "One of the huge successes was getting Mike Werner to be project scientist," Gehrz remembers. "Mike really facilitated this project by letting everybody do the things they did best and encouraging their participation. He was very even-handed and diplomatic…."[21] Werner pulled in people with diverse interests and motivations, whether they were on the Science Working Group or contractors developing the technology for SIRTF. He kept them involved (and focused) by emphasizing that their contributions would one day help to answer important scientific questions about the origins of the universe. SIRTF was needed to get the data, and obtaining it was Werner's prime motive for leading the project. Looking back on what has made Werner effective for nearly 30 years as project scientist, George Rieke observes:

> It's very simple. He was totally honorable. That means that whatever Mike told you, you knew he was telling you the truth. And that was really important…. [T]he fact that SIRTF was spread over many different institutions [required an] honest broker, to convince people that what was being done, even though it was very unpleasant for them, was being done for the good of the project, and in the long run it would be OK. It had to be somebody you had a huge amount of trust in, and Mike did that.[22]

Getting to Work

As project scientist, Werner presided over the meetings of the Science Working Group (SWG),

---

18. Werner interview (Lippincott), 25 July 2008.
19. D. Brandshaft, R. A. McLaren, and M. W. Werner, "Spectroscopy of the Orion Nebula from 80 to 135 microns," *Astrophysical Journal* 199, no. 2 (1975): L115–L117. M. W. Werner, I. Gatley, E. E. Becklin, D. A. Harper, R. F. Loewenstein, C. M. Telesco, and H. A. Thronson. "One arc-minute resolution maps of the Orion Nebula at 20, 50, and 100 microns," *Astrophysical Journal* 204, no 3 (1976): 420–423.
20. Werner interview (Lippincott), 25 July 2008.
21. Gehrz interview, 20 January 2009.
22. Rieke interview, 9 June 2009.

FIGURE 5.1. Participants in the first meeting of the Scientific Working Group, held at Ames Research Center in September 1984.

the main body responsible for specifying the design of SIRTF. With the selection process for the Announcement of Opportunity complete, Werner chaired the kickoff meeting of the SWG, which was held at Ames from 12 to 14 September 1984. Members of the SWG included Werner, Witteborn, and Nancy Boggess, as well as the instrument PIs—Giovanni Fazio, Jim Houck, and George Rieke. The SWG also included Frank Low, Mike Jura, and Ned Wright, each of whom had broad responsibilities (see Fig. 5.1). Attendees regularly included managers from Headquarters, SIRTF project office staff, and external engineering contractors.[23]

By the September 1984 meeting, NASA had already decided that SIRTF was going to be a free flyer and that the SWG would abandon the sortie option. The previous design had assumed that SIRTF would be mounted on a Spacelab pallet in the Shuttle's cargo bay and stay in space for a week or two. Instead, the Shuttle would now hoist SIRTF into space and leave it there. This change to a free flyer meant that the engineering studies had to be redone, a new design developed, and an appropriate orbit selected.

It would not be the last time that the SWG would select a new orbit. This time, however, the options were limited to a so-called low-Earth orbit, 900 kilometers up. This orbit would put SIRTF within reach of the Shuttle for servicing. With a free flyer, the SWG no longer needed to worry about the heat, dust, and water vapor in the Shuttle affecting SIRTF. However, the low-Earth orbit was still fraught with obstacles to observations—namely, Earth, the Moon, and the Sun. Other concerns included cosmic rays and

---

23. SIRTF Science Working Group (SWG) Meeting Minutes, 12–14 September 1984.

the South Atlantic Anomaly (SAA), which is a particularly low and intense region of radiation where the inner Van Allen radiation belt dips toward Earth. Even a brief glimpse of one of these warm objects can momentarily blind SIRTF's infrared detectors and quickly deplete the cryogen that keeps them cold.

To select an orbit that avoided most of these problems, the SWG worked with the SIRTF Study Office at Ames to compare an equatorial orbit (at a 28.5° angle to the Equator) with a polar orbit (at a 99° angle).[24] Regarding science objectives, a polar orbit would provide more observation time at longer wavelengths (~200 microns), whereas the equatorial orbit would allow for observations to be made simultaneously from SIRTF and the ground. Regarding technical performance, neither angle would entirely avoid the SAA. The equatorial orbit would avoid the SAA on many rotations, but when it passed through the Anomaly, the radiation would be intense enough that SIRTF would need to be shut down to protect the detectors (a strategy used with Hubble). In contrast, the polar orbit would take SIRTF through a less dense part of the SAA. However, it would do so on almost every rotation; so radiation would accrue over time, causing SIRTF's sensors to slowly degrade.

The equatorial orbit was unanimously adopted by the SWG in March 1985, although this "best" option allowed for only a paltry 24 minutes of observing time during each 90-minute orbit.[25] In between, SIRTF would need to dodge the heat of the Sun, the Moon, and Earth. During each orbit, SIRTF would need to be repositioned with enough time to settle before it could lock onto a new target. This would involve rolling, slewing, pointing, and nodding—motions that would be time-consuming and nontrivial for a telescope moving above Earth at 27,000 miles per hour.

Assuming these technical challenges could be overcome, the SWG was planning for SIRTF to have a two-year mission lifespan and to be capable of detecting infrared wavelengths from 1.8 to 700 microns, which is nearly the entire infrared spectrum. It was becoming clear that the limiting design factor would be the detectors, particularly for Fazio's and Rieke's instruments (IRAC and MIPS, respectively). So Fazio and Rieke agreed to jointly develop the detectors and try two approaches: repurposing existing military detectors that had been designed for the near-infrared and conducting basic research to develop detectors for the far-infrared.

Leading the detector development was Dr. Craig McCreight, a member of Fazio's instrument team. McCreight explains:

> What we were trying to do was to operate detectors at a much lower temperature than had been done before. These cryogenic detectors have to be quite cool to work at all, but we were going to take it even colder, because the cryogen tank was basically designed for around 2 K near the lambda point of liquid helium. And the key question was, Could these very sensitive detectors be pushed to lower temperature to fit the mission needs? The other technical thing we were trying to do, which was new, was to have [the detectors] stare for a very long period of time: to look at the sky, integrate the signal for quite a long period—as opposed to the military, which was operating really quickly, looking for fast-moving objects and things that we

---

24. The early members of the SIRTF Study Office included Walt Brooks (telescope facility manager), Silvia Cox, Ann Dinger, J. Givens, Bob Jackson, Larry Manning (systems engineering manager), Joe Mansfield, Ramsey Melugin, Jim Murphy (chief/project manager), Ken Nishioka, Rubin Ramos (science instruments manager), Gary Thorley (deputy project manager), and Chris Wiltsee.

25. For details on orbits, see *http://earthobservatory.nasa.gov/Features/OrbitsCatalog/printall.php* (accessed 30 August 2016).

often weren't able to talk very freely about. So it was a different application. We wanted to slow [the detectors] down and cool them down and see if they could approach what we needed.[26]

In addition to his role on Fazio's team, McCreight managed Ames's advanced detector program and had helped to develop the detectors for IRAS, which used 62 discrete detectors that operated between 12 and 100 microns. Although these were not integrated arrays (the chip with 62 detectors was about the size of a coffee cup), they nevertheless provided a matrix with which "you could image the sky," McCreight said, "as opposed to having to move your 'little' single, discrete point detector and raster it around. And the fact that you could stare at an area of the sky and not move the telescope greatly simplified the pointing requirements of the telescope. There are probably a range of systems or requirements on the telescope that got eased or relaxed or canceled because of the power of the staring array. It was pretty important."[27]

One Step Forward, Two Steps Back

To develop the detectors and other components of SIRTF, the project would need much more funding. By the time the SWG was formed, approximately $15 million had been spent on SIRTF, or a little more than $1 million a year since Fred Witteborn had started working on the project in 1971. SIRTF needed a "New Start," a formal request to Congress for the funds needed for NASA to build and launch a major project. This process requires project approval from NASA Headquarters and budget approval from Congress. When the SWG first met, in September 1984, they projected that SIRTF would obtain a New Start in FY 1987. By December, that prediction had changed to FY 1989, because another project had jumped ahead in line. Although no one knew it at the time, SIRTF would not get a New Start until FY 1998. From the inception of the project, schedule after schedule showed the New Start as always at least three years out, with launch three years after that.[28] This perpetually revised schedule was humorously represented as a Mobius strip by project manager Lou Young (Fig. 5.2).

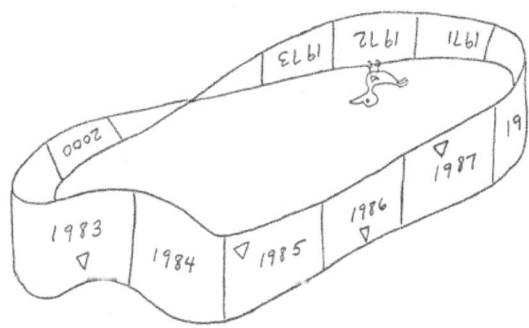

FIGURE 5.2. A humorous take on the SIRTF project schedule, with the New Start always in the future (*SIRTF Coloring Book*).

---

26. Craig McCreight, interview by author, Mountain View, CA, 2 September 2008.

27. McCreight interview, 2 September 2008.

28. For example, as of October 1979, the New Start was expected in FY 1982, with launch in FY 1996, per the Phase A Statement of Work (SOW); see Larry Manning, "Space Infrared Telescope Facility. Statement of Work Specification, Phase A," AFS1070.8A, presentation dated 4 October 1979 and housed in the NASA Ames History Office Archives Reference Collection, FC5: D4, Ames Research Center, Moffett Field, CA. By contrast, in February 1989 it was expected that a New Start would occur in FY 1992, with launch in FY98; see *Proceedings of the Third Infrared-Detector Technology Workshop*, comp. Craig R. McCreight, ARC-12851 (Moffett Field, CA: NASA, 1989); NTRS 19900011997.

The process by which projects at NASA obtain a New Start is revealed in the stories of how SIRTF became a member of NASA's Great Observatories and how another project, the Advanced X-ray Astrophysics Facility (AXAF), gained a New Start before SIRTF did.

The ordering of NASA missions is anything but straightforward. Qualifying as a New Start depends on whether the technology is feasible, the budgets are reasonable, and the mission fits with both the nation's policy objectives and other missions already committed to by NASA. All of these criteria are like stakes driven into quicksand, because anything and everything can change—officeholders leave, technology becomes obsolete, and economic conditions shift. Given the complexity of prioritizing missions in a changing environment, advocates must be absolutely committed to a project and skilled at "selling it."

For astronomy missions, the person who sat at the intersection of competing political, economic, and scientific priorities was NASA's Director of Astrophysics.[29] This person, like other divisional directors in NASA's Office of Space Science and Applications (OSSA), was not a political appointee and might stay in the role long enough to seize (or create) the opportunity for a New Start.[30] When Frank Martin was in this role, he single-mindedly pursued IRAS, which launched under his directorship. However, Martin also made sure to keep missions across the electromagnetic spectrum on life support with small pockets of money from general development and discretionary funds, and he also ensured that COBE (launched in 1989) received a New Start.

Charlie Pellerin, who replaced Martin in 1983, also had a single-minded focus. Under Pellerin's tenure, AXAF received a New Start. However, Pellerin did not particularly care if AXAF or SIRTF went first. What he was determined to accomplish was to put a suite of telescopes into space that could, for the first time in history, observe objects simultaneously across the entire spectrum. Although Martin, his predecessor, had laid the foundation, it was Pellerin who would bundle together the four major telescope programs that spanned the spectrum and were in various stages of development: Hubble, AXAF, SIRTF, and the Compton Gamma-Ray Observatory (CGRO). Although separate scientific communities championed each project, referring to the telescopes collectively as "the Great Observatories" made it much easier to sell all of them.[31]

Joining the Great Observatories

The majority of NASA's budget in the 1980s was being spent on Hubble, the Space Shuttle, and planning for a crewed space station; and cost overruns under Martin and Pellerin were further reducing the money that was left for other projects. All such projects were one of a kind and state of the art, so it was not surprising that cost estimates were inaccurate. IRAS came in $60 million over its initial budget. By the time Pellerin took over the division, Hubble had overrun its budget by $400 million, and millions more would be needed to complete it. (After Hubble launched on 24 April 1990, even more funding would be required to fix the spherical aberration of its primary mirror.) The project members of AXAF and SIRTF were continually badgering Pellerin for a New Start, but resources could support only one. No matter which project he picked, he would have a fight on his hands. As Pellerin recalls:

---

29. NASA's Astrophysics Division is now known as the Solar and Terrestrial Astrophysics Division.
30. NASA's Office of Space Science and Applications (OSSA) is now known as the Science Mission Directorate.
31. Pellerin interview, 19 March 2009.

I might not be able to sustain it even as Director of Astrophysics. They would go over my head. They would go to the decadal survey [committee], they would go to my bosses on the Hill. They would fight me everywhere.... What I've got to do is come up with a story that gets everyone to support the whole program. I've got to get them to believe in me. Some tale that's going to get everybody behind everything.... I need help for Hubble. It's overrunning all the time. It's not going to get canceled, but if I don't have political support, it's just going to be more difficult. The gamma-ray observatory is very fragile, because it's a very small community of people. The number of people doing gamma-ray astronomy is minuscule, and it's vulnerable [to cancellation] at any moment.... And then I've got SIRTF and AXAF as the two movable chips. So what I had to do was get people to believe in me when I told them that we were going to do all four [Great Observatories].... That was the only way I could succeed politically with this program.... I said, "What's the physics story here?"... About that time, a guy named Dave Gilman, who worked for me, either told me about or gave me a copy of Martin Harwit's book called *Cosmic Discovery*. Frank [Martin] had heavily relied on that book, too. I didn't know that until later....

I was trying to get data of a sort that never existed before. We never had the measurements of an object across all the things that were going on in it.... So I looked in Martin's book.... [T]his is the ammunition for making the discovery argument [through] the advancement of capabilities. And I can show, for each of these missions—I'm not talking about astronomy, I'm talking about the advancement in measurement capability, the historical trend curve and the discoveries that have come out of each one. The fact that all together they can bring physics that we have never seen before. That was the sales pitch...to everybody. Now, my biggest problem was not the science community in getting it going forward; my biggest problem were the bureaucrats in Washington who controlled the money: the staffers on the Hill, the people in OMB [Office of Management and Budget]. I talked to them early on and I recognized that when I told them I was the Director of Astrophysics, their eyes glazed over, because what they believed was that they couldn't understand it. None of them had technical degrees. They all remembered flunking high-school physics or something. So what they're saying to themselves is "Whatever he's going to say is beyond my capacity to understand." So I'm pondering this, and I'm asking myself, "What's the least intimidating medium I can use to communicate this, because I've got to get this message across. How about a comic book?"[32]

Pellerin asked his deputy, George Newton, to invite some of the top astrophysicists to come to Washington on 3 January 1985 and draw cartoons that could be used to communicate the benefits of this four-telescope program. Among those who accepted the invitation were key scientists for the Hubble, AXAF, SIRTF, and CGRO, as well as members of past decadal surveys. It was a prestigious group. "I started to think about this meeting, and I began getting nervous," Pellerin said. "When I was a practicing scientist, these guys were God. I knew I was smarter than they were about what we were doing here, but I wanted to

---

32. Pellerin interview, 19 March 2009.

make sure I could show them that.... To do that, I'd have to have someone else run the meeting. So I thought, Martin Harwit wrote this book—why don't I get Martin Harwit to run this meeting? I'd never met him, so I called Dave Gilman [and Dave said], 'He's a great guy.'"[33] Encouraged by Gilman's recommendation, Pellerin then called Harwit, who was rather surprised by the request but readily agreed to chair the meeting.[34]

At the meeting, Pellerin and Harwit explained to the other scientists the context of the day's meeting. They drew on the list of cosmological questions that Harwit laid out in his book—How are stars born? How do quasars work? "We took flip-chart paper and we put a question on top," Pellerin recalls. "And then I got the top astrophysicists in America on their hands and knees down there, using these crayons."[35]

Harwit took the drawings and, with around $25,000 from Pellerin, turned them into a publishable booklet. He was assisted in this task to make the science understandable by Valerie Neal of the Essex Corporation, who had a doctorate in the humanities, and by illustrator Brien O'Brien. Pellerin had 15,000 copies printed and distributed them widely. "They came in on a forklift," Pellerin remembers, "and we took them on the Hill. We took them to schools. We did everything except drop them out of airplanes....We [went] to everybody that matters and they [got] to hear the story: a new physics in our lifetime—American leadership through these observatories.... In a year, the Great Observatories was a household word."[36]

Prioritizing the Missions

Having conditioned Congress and the OMB to welcome the Great Observatories concept, Pellerin's next step was to get NASA to authorize a New Start for either AXAF or SIRTF. It did not matter to Pellerin which one was funded—both were part of the Great Observatories. "I asked myself, what's the optimum sequence to get them all done as quickly as possible?" Pellerin said. "AXAF had contracted studies, SIRTF didn't.... We had Marshall spending a lot of money [on AXAF]. The design was very mature. I had the decadal survey supporting me. I had all these things. AXAF was just further along [than SIRTF]. I wasn't interested in which one went first, I was interested in how fast I could do all of them."[37]

People close to SIRTF might disagree that their project was not as mature as AXAF. The Field report had indeed given AXAF top priority for new missions. However, it also explicitly characterized SIRTF as an ongoing mission, along

---

33. Pellerin interview, 19 March 2009. Participants in the comic book development meeting included Robert A. Brown, from Marshall; Carl Fichtel and George Pieper from Goddard; George Field, Josh Grindlay, Robert Noyes, Irwin Shapiro, and Harvey Tananbaum, all from the Harvard-Smithsonian Center for Astrophysics; Riccardo Giacconi from the Space Telescope Science Institute; Bill Hoffmann and George Rieke, from the University of Arizona; Ken Kellermann from the National Radio Astronomy Observatory; Jeremiah Ostriker from Princeton; Edwin Salpeter from Cornell; and Rainer Weiss from MIT. For this list and a detailed discussion of the Great Observatories based on Harwit's papers in the Cornell archives, see Martin Harwit, "Conceiving and Marketing NASA's Great Observatories," Experimental Astronomy 26, nos. 1–3 (2009): 163–177.

34. Martin O. Harwit interview, Cambridge, MA, 26 May 2009.

35. Pellerin interview, 19 March 2009.

36. Pellerin interview, 19 March 2009; see also Harwit, "Conceiving and Marketing NASA's Great Observatories." Two different versions of the comic book were produced, both titled "The Great Observatories for Space Astrophysics." The first (NASA-CR-176754/NASA NTRS Doc 19860015241) was printed in 1985; the second version (NASA-NP-128 / NTRS Doc 19920001848) was printed in 1991.

37. Pellerin interview, 19 March 2009.

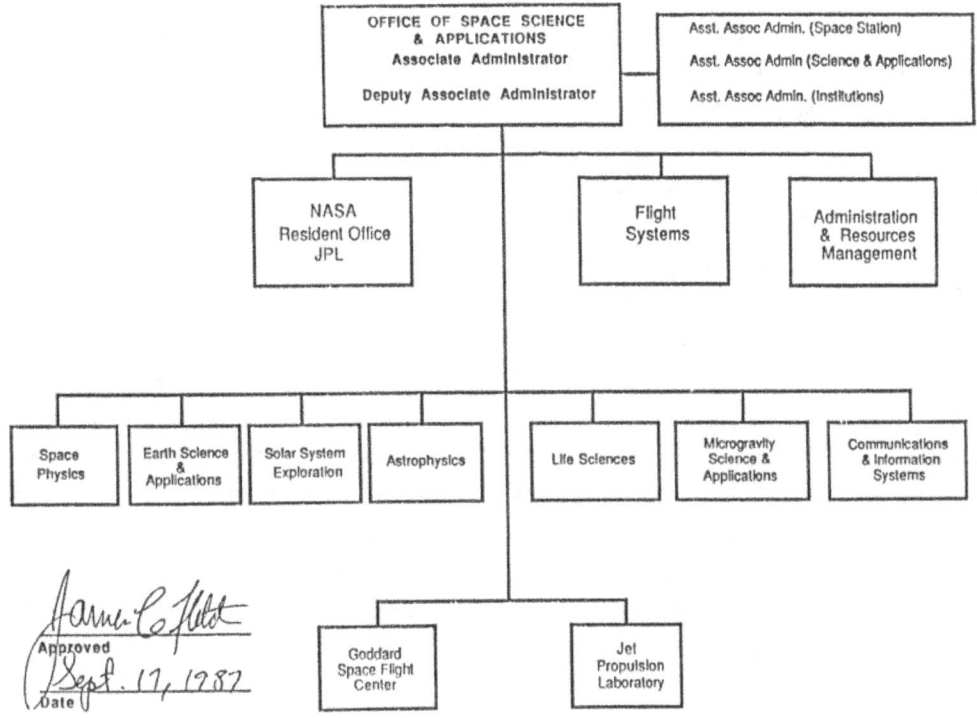

FIGURE 5.3. Organizational chart of NASA's Office of Space Science and Applications (September 1987 SIRTF SWG Meeting Minutes).

with Hubble and CGRO.[38] The key difference between the projects was that SIRTF did not yet have a New Start, while Hubble (FY 1978) and CGRO (FY 1980) did.[39] This put SIRTF on less-certain ground. Moreover, the Field report had assumed that SIRTF would be a Shuttle-attached payload. Now SIRTF, like AXAF, was being planned as a Shuttle-launched free flyer. AXAF had been on this development path since the beginning, whereas SIRTF had switched to being a free flyer in 1984. Pellerin estimated that it might take two years for SIRTF to catch up with AXAF.[40] Although the technologies required for SIRTF had already been largely demonstrated by IRAS, SIRTF was in the middle of a major redesign. AXAF was not. Others have characterized this episode as a competition between AXAF and SIRTF,[41] but the reality is less dramatic. One telescope had to go first. Pellerin simply picked the one that was most sellable at that moment.

Selling SIRTF internally at NASA was still a challenge. Pellerin's Astrophysics Division was just one group within the Office of Space Science and Applications (see Fig. 5.3), and each such group

---

38. Field report, vol. 1, pp. 13–14.

39. Gamma-Ray Observatory Science Working Team, "The Gamma-Ray Observatory Science Plan, September 1981," in Logsdon et al., eds., *Exploring the Unknown: Selected Documents in the History of the U.S. Civil Space Program, Volume 5: Exploring the Cosmos*, pp. 692–697, available at http://history.nasa.gov/SP-4407/vol5/chapter-3/III-32%20%28292%29.pdf.

40. Pellerin interview, 19 March 2009.

41. This is a theme in the history of AXAF presented by Wallace Tucker and Karen Tucker in *Revealing the Universe: The Making of the Chandra X-ray Observatory* (Cambridge, MA: Harvard University Press, 2001).

had a set of projects it wanted to move forward. "There was a lot of in-fighting about which New Start was going to happen and when," Pellerin said.[42] It fell to Burton I. (Burt) Edelson, NASA's Associate Administrator for Space Science and Applications (1982–1986), to make a recommendation. Edelson managed by consensus, or at least aspired to, despite the conflicting demands and budgetary constraints that caused fierce competition among the NASA Centers and divisions.[43]

To select the New Starts for the near future, Edelson brought his staff and other interested parties together at the Belmont Conference Center in Elkridge, Maryland, for a weekend planning retreat. "It was a very unfair setting for me," Pellerin recalls, "because Burt had invited people who were opposed to my program, like the director of the Jet Propulsion Lab, Lew Allen. [He's] a great man, but if he was going to be there, the Director of Marshall should have been there, too, and he wasn't invited; so it was a stacked deck. These people were going to vote on something they knew nothing about. There was no opportunity to educate anybody."[44] Even with the help of a professional facilitator, by the last day the attendees had still not addressed the New Starts. As Pellerin recalls it,

> I went to the facilitator and said, "We came here to do this [New Start selection]. When are we going to do it?".... So they put up these viewgraphs, and they had these projects in some order. There were like twelve candidates for New Starts [and] AXAF was low on the list.... We'd have a sequence of the top four and then the rest would go away. I had two or three up there. I had AXAF. SIRTF wasn't on the chart then, I don't think.... So the facilitator said, "Let's take a straw vote to see how we all feel about this." I'm sitting there and I'm thinking, if we vote, it's over. Burt is not going to have a straw vote and then continue to discuss this. So we go around the table one at a time, and everyone was being very politic: three votes for this, four votes for this. There were twelve candidates, and we each had twenty-four votes [to distribute among the projects].... Before we started voting, I asked the facilitator, "What are the rules on voting?" He said, "It's just a straw vote, so you can vote any way you want." He goes around the table, and I'm the next to the last person, and I'm looking at this, and AXAF is going no place. So I said, "Twenty-four votes for AXAF." Burt jumps up and screams at me, "You can't do that!" I said to the facilitator, "I asked you about the rules, and you said the rules say I can." He looks at Burt and says, "He's right, that's what I told him the rules were." Burt was steaming mad, because he thought he'd managed to get rid of AXAF. And the person after me did it political again: two for this, one for that, four, two. So as a result, AXAF came in second out of the twelve. That's the only year in NASA I didn't get a bonus, because Burt was so mad about that....
>
> I got my project. Right after the top planetary mission.... That's how it was decided. And exactly what I knew would happen is that Burt would collect up all the charts and we'd leave. I knew we'd never have a

---

42. Pellerin interview, 19 March 2009.

43. Edelson's preference for consensus is remarked on both by Pellerin (19 March 2009 interview) and by Harwit ("Conceiving and Marketing NASA's Great Observatories," p. 170). Edelson, who died in 2002, was a longtime public servant, having served in the Navy (1943–1967), at NASA (1982–1986), and, after his retirement, at George Washington University's Institute for Applied Space Research; see *http://www.arlingtoncemetery.net/biedelson.htm* (accessed 30 August 2016).

44. Pellerin interview, 19 March 2009.

conversation after the straw vote, because I knew Burt. Burt was uncomfortable with any conflict; he couldn't stand it. This was the biggest conflict: Whose billion-dollar thing comes next? And to Lew Allen's great credit, he got up in the middle of it and said, "You know, it's a really clear choice here. If you want to support JPL institutionally, you're going to vote for my mission. If you want the best science, you're going to vote for Charlie's." I never quit respecting that bit of integrity from him.[45]

This episode illustrates how opportunities to "sell" a project do not always appear in a predictable way and how politics and personal idiosyncrasies can shape the outcomes. However, this is not to say that such decisions are made irrationally. Every project being considered for a New Start was thoughtfully vetted by dozens of smart, creative, skilled scientists and policy makers. Every project had a solid rationale for being funded. Unfortunately, there were more good ideas than money to develop them.

The Federal Budget

With Burt Edelson committed to supporting AXAF as a New Start, Pellerin was one step closer to realizing the Great Observatories concept. The next hurdle was getting Congress and President Reagan to sign a budget that authorized spending on AXAF, the next of the Great Observatories.

Every spring, the Office of Management and Budget (OMB) starts working with federal agencies to develop a budget that the President will present to Congress. The OMB provides financial guidance as to how much funding each agency might expect and tries to align the budget with the President's goals while balancing the demands of health, defense, and national priorities.[46] The President's budget, traditionally presented on the first Monday in February, offers recommendations as to how discretionary monies should be spent to meet policy goals in the upcoming fiscal year (beginning 1 October). Discretionary funding makes up nearly a third of the national budget and affects nearly everything with an acronym: FDA, EPA, DOD, HUD, NIH, and many others, including NASA.

Congress then takes these recommendations and marks up the budget after hearing from the directors of the various agencies, who are brought before Congress to defend their budgets and project priorities. Both houses of Congress modify the budget and come to an agreement on overall fiscal policy that is (ideally, but not always) in line with the President's recommendations. This policy document is not legally binding but serves as the basis for legislation in which funds are actually granted to various agencies and projects.

Every year, the budget process is repeated and often begins before the current budget is finalized (see Fig. 5.4, p. 78). This federal budgeting process makes it difficult for NASA to secure funding, as all long-term projects require annual approval, and budget requests must be submitted nearly two years in advance, regardless of changing national priorities and development costs.

---

45. In the account of AXAF's history given by Tucker and Tucker, the authors reference a meeting of NASA's *Space and Earth Science Advisory Committee (SESAC)* in which two programs being managed by JPL were ranked higher than AXAF: the ocean Topography Experiment (*TOPEX*), an Earth ocean mapper) and the Comet Rendezvous Asteroid Flyby (CRAF), one of the two flagship missions of the Mariner Mark II program (*Revealing the Universe*, p. 94). CRAF was canceled in 1992, while Cassini was launched in 1997.

46. For a more detailed description of the federal budget process, see "Introduction to the Federal Budget Process," Report no. 98-721, 3 December 2012, by Bill Heniff, Jr., Megan Lynch, and Jessica Tollestrup (Washington, DC: Congressional Research Service, 2012).

The Executive Budget Process Timetable

Date	Activities
Calendar Year Prior to the Year in Which Fiscal Year Begins	
Spring	OMB issues planning guidance to executive agencies for the budget beginning October 1 of the following year.
Spring and Summer	Agencies begin development of budget requests.
September	Agencies submit initial budget requests to OMB.
October–November	President based on recommendations by the OMB director makes decisions on agency requests. OMB informs agencies of decisions, commonly referred to as OMB "passback."
December	Agencies may appeal these decisions to the OMB director and in some cases directly to the President.
Calendar Year in Which Fiscal Year Begins	
By first Monday in February	President submits budget to Congress.
February–September	Congressional phase. Agencies interact with Congress, justifying and explaining President's budget.
By July 15	President submits mid-session review to Congress.
August 21 (or within 10 days after approval of a spending bill)	OMB apportions available funds to agencies by time period, program, project, or activity.
October 1	Fiscal year begins.
Calendar Years in Which Fiscal Year Begins and Ends	
October–September	Agencies make allotments, obligate funds, conduct activities, and request supplemental appropriations, if necessary. President may propose supplemental appropriations and impoundments (i.e., deferrals and rescissions) to Congress.
September 30	Fiscal year ends.

Source: Office of Management and Budget. Circular No. A-11 (Washington: July 2007). Section 10.5.

FIGURE 5.4. Timeline of the federal budget process.

Appropriations are as much a financial process as a political one. Thus, it is not surprising that Congress, the President, and the OMB can and often do decide the fate of individual projects, particularly if those projects are expensive or politically visible. For example, Cold War concerns led President Kennedy to push the Apollo program and President Reagan to push the Star Wars defense system despite the objections of scientists and engineers who thought the respective technologies were not yet possible.

NASA's New Starts must therefore have broad political and fiscal appeal. But it is a challenge to align the interests of presidents, congressional representatives, OMB staff, and agency directors and keep them aligned throughout an 18-month budget cycle. That is a lot of time for something to go wrong.

Appealing to Power

Charlie Pellerin was pleased that the NASA budget submitted to the OMB in mid-1987 included at least one of the Great Observatories, AXAF. However, the two preceding years had been hard. NASA had been consumed by the congressional investigation of the Challenger explosion in January 1986, in which the Shuttle's seven crewmembers had been killed. At the time, NASA was without a permanent Administrator.

The outgoing Administrator, James M. Beggs, had been on leave since December 1985; and his successor, James C. Fletcher, would not be confirmed until the following May. Changes were also happening in the Office of Space Science and Applications: Burt Edelson had retired, and Lennard A. Fisk had become the Associate Administrator for OSSA in May 1987.[47] About the only thing that was unchanged was the cost overruns on Hubble. Thus, when the budget was submitted, Pellerin had two new bosses and regard for NASA was at an all-time low.

Although Pellerin's division was not directly involved in the Shuttle investigations, no telescope was going to fly on a Shuttle for the time being. If anything was going into space, then the first in line was Hubble, which was now in a holding pattern and burning tens of millions of dollars as it kept teams together waiting for their chance at launch. AXAF was next, but it wasn't going anywhere until Hubble was back on track. And SIRTF would have to find a new ride, because the supercooled helium cryogen it relied on was highly pressurized—and highly explosive—a risk that was now seen as too great to take on a Shuttle.

Although Fisk saw to it that AXAF made it into NASA's FY 1988 budget, it was dropped in January 1988, just days before President Reagan was to give his State of the Union address.[48] Pellerin had to act fast and relied on a process called *reclama* to try to reverse that decision. "'*Reclama*' means that you go to the [NASA] administrator and he agrees to go back and appeal this," Pellerin explains. "The way the appeal works is that if it's a big thing like this, you go and you make a presentation to the President in person, and you try to overrule his OMB director. So the OMB director is going to negotiate with you to avoid this, but he's not going to negotiate much, because he's the President's man."[49]

Pellerin first needed to get the support of the new NASA Administrator, Jim Fletcher. Fortunately for Pellerin, he had a very good relationship with Fletcher, who had previously served in this role (1971–1977) under Presidents Nixon and Ford. After Fletcher left office in 1977, he joined University Corporation for Atmospheric Research (UCAR), a consortium that managed NASA's balloon program, which is where he met Pellerin. "Most people, after Fletcher left [NASA], treated him like a nobody," Pellerin recollects.

> I always thought that since he had been the administrator, I was going to give him the respect an administrator gets, even though he's not the administrator anymore. I just thought it was the decent thing to do, to treat him the same way. So I had meetings with him and met with his board, and I was deferential to him.… So now he's back [at NASA] and I've got this relationship with him that I built during the years he was gone.… So when AXAF got taken out of the OMB budget, I went to meet him in person, with Fisk, and I said, "Jim, you've got to put this back in." He said, "No, Charlie, I'd like to help you, but they've cut manpower costs, they've cut the Shuttle main-engine testing, they've cut the facilities. NASA's got more important things than your science project." I said, "Let me explain what this is, Jim. This is the Great Observatories program."[50]

---

47. See Lennard A. Fisk, interview by Rebecca Wright, Ann Arbor, MI, 8 September 2010, at *http://www.jsc.nasa.gov/history/oral_histories/NASA_HQ/Administrators/FiskLA/FiskLA_9-8-10.htm* (accessed 30 August 2016).

48. Pellerin interview, 19 March 2009; for a more complete account of AXAF's history, see Tucker and Tucker, *Revealing the Universe*.

49. Pellerin interview, 19 March 2009.

50. Ibid.

Fletcher authorized Pellerin to make a case to take before the OMB and, ultimately, President Reagan. With only a few days before Reagan would unveil his budget, there was no time for a letter-writing campaign, a blue-ribbon panel, or any of the other strategies that AXAF's advocates had become adept at for mustering support.

What would persuade Reagan? The Challenger explosion had happened on his watch, two years earlier. Since that time, the Soviets had put into orbit the Mir space station and the Kvant-1, a module for x-ray and optical astrophysics, which had launched and successfully docked with Mir the previous spring. The only thing they had not done was fly their own piloted space-transport system, but they were preparing to, and their vehicle looked nearly identical to the U.S. Space Shuttle. Even so, it wasn't all competition with the Soviets: Pellerin was involved in building a cooperative program with the Soviet Union in space (paralleling efforts in the Manned Space Directorate, which managed a 1995 mission in which the Shuttle docked with Mir). Pellerin had become good friends with his Soviet counterpart, Rashid Alievich Sunyaev, who had given him a picture of Mir with Kvant-1 attached to it.[51]

It occurred to Pellerin that he might be able to use this image of Mir to persuade the OMB director and President Reagan that AXAF needed to be funded. Working with Trish Pengra of BDM Corporation, a professional services firm that provided regular support to the Astrophysics Division, Pellerin developed a compelling (and mostly truthful) story. It was about the Russians: "We put a picture of an American flag you could see through with AXAF and a Soviet flag you could see through with Mir," Pellerin said. "And of course, I didn't label that little chunk on the end as the x-ray part. They wouldn't know, right? And I put, 'To whom will the future in space astronomy belong?' as the title. Now, if there were things more important than AXAF to Mr. Jim Fletcher, is this ever going to get out of NASA this way? No. There's no way." Pellerin understood that since this image could be seen by the President, everyone who ranked between Pellerin and the President would want to review the presentation. That would take too long, and someone was sure to come along who wouldn't like it. "I had to make a choice," Pellerin said:

> I decided that I would see what I could get away with and I called Fisk's administrative assistant and said, "I've got some charts for OMB, does anybody want to see them?" The assistant said no. Then I called Fletcher's office. I got some administrative person, and I said, "I've got some charts for OMB."… They had no idea what charts I was talking about. I made it sound real casual. They must have thought just some routine stuff.… Trish found out that the President doesn't hurt his eyes. He only looks at [2- by 4-foot] foam boards. So Trish finds out who [makes those] and has four or five slabs made up, and the last one is this slab. So we got the foam boards. I wrapped them up in a brown wrapper. I put a NASA sticker on there with no cover letter, and I just put "Director, OMB," with no return address. We hired a courier to take it to the OMB director's office, [and he] drops this package from NASA off. The next day I get a call from Fisk and he says, "Fletcher just called me and said that the damndest thing happened. AXAF is back in the budget. Do you know anything about it?" "No sir.…" I mean, let's be honest, I pushed higher priority things to the Agency aside for this. If I got caught doing this, I might be fired.… Nobody figured it out. [Years later] I had a guy in a workshop at Huntsville who worked

---

51. Ibid.

at OMB at the time. He said this package came in and it created this enormous stir. He said no one had ever seen anything like this before. The OMB director looked at it and said, "Tell me again, why did we take this out?" It turns out that some lower-level guy had just thrown AXAF out for the hell of it; there was no clear rationale as to why they'd done it. The OMB director said, "I'm going to go to the President with this chart? Put that thing back in." And that was the end of it.[52]

The story of how AXAF moved ahead of SIRTF is not about which project was better. They were both important to Pellerin and the scientific community. AXAF did not launch first because its advocates were more persuasive but because there was not enough money to fund both and one had to be first. SIRTF had made a design change to a free flyer that was substantively different from what the Field committee had endorsed. New engineering work for SIRTF to be a free flyer was needed, whereas AXAF had always been on that track.

This a story about how careful scientific and technical work, coupled with seemingly random processes at the economic and political levels, created opportunities for projects to get on or get off track.[53] Seizing such opportunities required initiative, such as Pellerin displayed, but also support for such actions. Len Fisk noted, "It was my management practice to give my division directors a lot of license, particularly when they were as ingenious as Charlie. They might have been surprised to know how much I actually knew of their actions, and found ways to encourage them."[54]

It is a story that describes many accounts of large-scale projects—not just AXAF and SIRTF. Its conclusion depends on whether project managers and budget-level administrators establish favorable circumstances for achieving their goals. Seemingly unrelated events or circumstances can sometimes propel this narrative: a book that influenced NASA policy-makers written only because Martin Harwit had spare time for writing after NASA cut his funding; a comic book that bundled together a set of independently conceived projects, making all of them easier to sell to Congress; the Shuttle disaster that ended once and for all any speculation that SIRTF (and its volatile pressurized-helium cryogen) would be an attached payload; and competition in space that spoke to the "Star Wars" president. All of these factors existed outside the project teams, where decisions were essentially divorced from technical and scientific considerations. Politics, economics, and personalities all can drive the prioritization of projects. When a mission like SIRTF gets off the ground, it is because many hands have been there to lift it, some pushing, some pulling, and some obstinately holding the tethers until it can no longer be kept down.

---

52. Pellerin interview, 19 March 2009. Pellerin's account was confirmed by Lennard Fisk and Trish Pengra (e-mail exchanges with the author, 10–12 October 2011).

53. Political scientists and policy scholars have developed ideas around policy windows and venues, as an insightful reviewer noted. For more detail, the interested reader is directed to John W. Kingdon, *Agendas, Alternatives, and Public Policies*, 2nd ed. (New York: Longman, 1995); Frank R. Baumgartner and Bryan D. Jones, *Agendas and Instability in American Politics*, 2nd ed. (Chicago: University of Chicago Press, 2009); and Bryan D. Jones and Frank R. Baumgartner, *The Politics of Attention: How Government Prioritizes Problems* (Chicago: University of Chicago Press, 2005).

54. Lennard Fisk, e-mail exchange with the author, 12 October 2011.

CHAPTER 6

Out of Step

The SIRTF team was disappointed to see AXAF move ahead of them to the front of the line. They were more than a little surprised, too. Members of both the SWG and the SIRTF project office at Ames had assumed not only that SIRTF would be the next major New Start but that they had put into place the supports necessary to justify this position—scientifically, politically, technologically, and economically.

Realigning Support for SIRTF

In the three-plus years since its first meeting, the SIRTF SWG had reached out to the scientific community to garner support. The most influential group of scientists was still the optical astronomers, for whom an infrared telescope had limited appeal. However, astronomers of all stripes were increasingly studying planets, which are relatively cold and dark objects and thus well suited to examination in the infrared. By the 1980s, planetary astronomy had shed the image associated a century earlier with Percival Lowell's description of Martian canals and the ensuing popular speculations about life on Mars. Over the preceding two decades, optical astronomy and planetary astronomy had gradually converged, in large part due to NASA's planetary missions, such as Mariner (which revealed the "canals" to be an optical illusion), Pioneer, and Voyager.

Planetary astronomers had become a powerful group, and their support for SIRTF was highly desired. To represent the interests of these scientists, Dale Cruikshank, of NASA Ames, was added to the SWG. Cruikshank's doctoral advisor at the University of Arizona had been Gerard Kuiper (see chapter 2), who advocated both infrared and planetary research before either was fashionable. As a member of the SWG, Cruikshank's role was to incorporate ideas on how SIRTF could best be used for planetary research.[1]

The SWG also designated one of its members, Mike Jura, as the political point man. Jura, who was one of the SWG's interdisciplinary scientists, initially focused on providing information about SIRTF to the advisory boards that shaped national science priorities and policy, such as NASA's Space and Earth Sciences Advisory Committee (SESAC).[2]

---

1. Cruikshank became advisor to the Science Working Group in August 1986 and was formally made an SWG member in October 1988 (SIRTF SWG Meeting Minutes, 12–14 August 1986 and 17–19 October 1988).

2. SIRTF SWG Meeting Minutes, 15–17 December 1986.

Other outreach efforts by the SWG (with help from the Ames project office) included a brochure on SIRTF for lay audiences and a newsletter to keep the scientific community informed of their progress. They also edited a special issue of the academic journal *Astrophysical Letters and Communications* that detailed the scientific capabilities of SIRTF as a free flyer.[3]

What the SWG failed to appreciate was the short shelf life of decadal surveys; nor did it anticipate that shifting SIRTF from a Shuttle-borne facility to a free flyer would make the relevant part of the Field report obsolete, and with it support for SIRTF. Although the move to a free flyer improved the observational quality of the telescope, it also made SIRTF seem like a brand-new project and therefore in need of reprioritizing in the next decadal survey.[4] The move to a higher orbit also introduced new technological considerations and tradeoffs that would require evaluation before a design could be specified.

In addition to this erosion of scientific and technical support, economic and political support was dwindling. Outside advisory groups had expressed growing concern with NASA's overall ability to reach its many goals. One particularly influential review was conducted for the White House by the Advisory Committee on the Future of the U.S. Space Program, chaired by Norman Augustine, who was then CEO of Martin Marietta. In its 17 December 1990 report, the so-called Augustine Committee recognized the progress of the Great Observatories program and other NASA initiatives (see Fig. 6.1) yet raised serious issues:

With so spectacular a set of achievements as a foundation, and with a substantial number of space projects underway, the U.S. space research enterprise should be healthy and flourishing. Yet discussions with researchers within NASA and in the university community reveal that there is significant discontent and unease about what the future may hold for U.S. space research. The reasons for these concerns have been documented in some detail in the 1986 report entitled "The Crisis in Space and Earth Science" issued by the NASA Advisory Council. They include such factors as (a) the widening of research horizons in response to past accomplishments so that there are now more opportunities than can be accommodated by the available resources; (b) the space technology required to support new advances is often more costly and sophisticated than in the past; (c) the growing complexity of interactions between NASA and its larger and more diverse research community; and (d) program stretch-outs, delays and cancellations that waste creative researchers' time, squander resources, and decrease flight opportunities. We believe that many of these reasons continue to exist.[5]

The problems identified by the Augustine Committee were largely due to a lack of resources—a problem for which there was no simple solution, as NASA's budget was being squeezed from two sides. On one side were several large and technically complex programs, such as the Shuttle and the International Space Station, that ate up the majority of the budget. Problems with the Shuttle required additional

---

3. See "Spacelab 2 Mission," a special issue of *Astrophysical Letters and Communications* 27, no. 3 (1988).
4. Robert D. Gehrz, interview, Long Beach, CA, 5 January 2009.
5. Advisory Committee on the Future of the U.S. Space Program (Norman Augustine, chair), *Report of the Advisory Committee on the Future of the U.S. Space Program* (Washington, DC: NASA, 1990), hereafter cited as the "Augustine report".

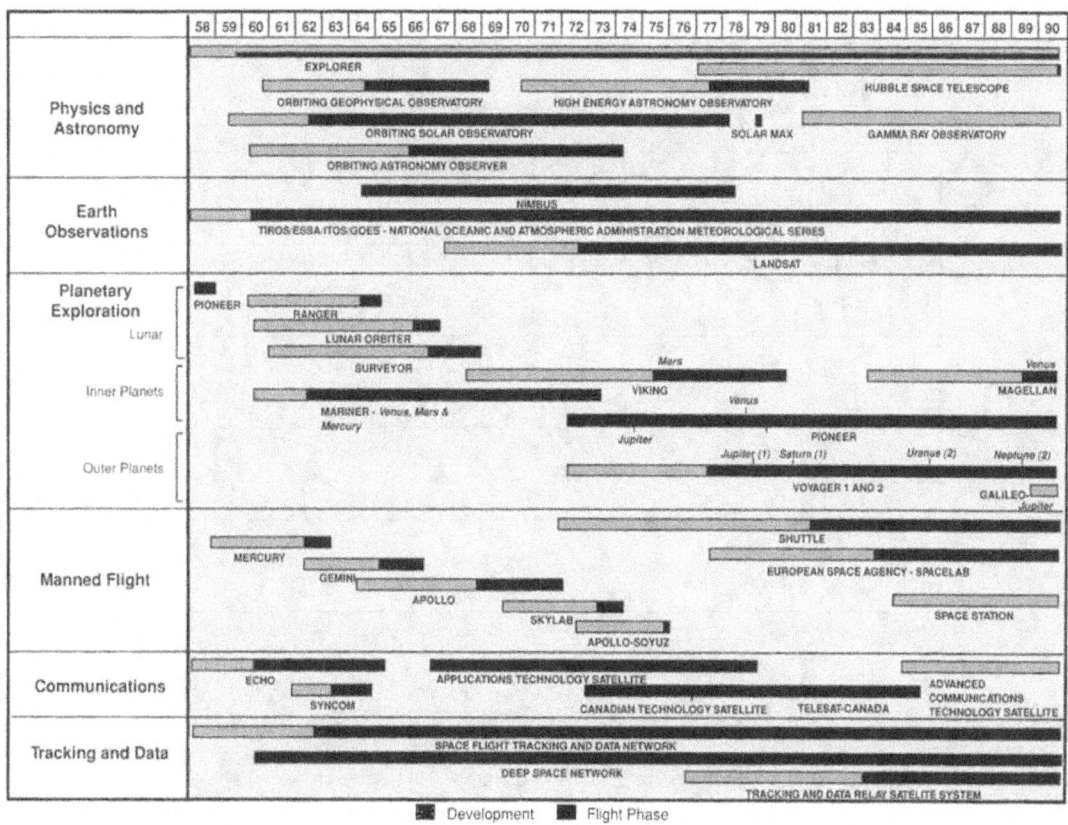

FIGURE 6.1. Major NASA projects from 1959 to 1990.

funding in the late 1980s and created overruns in the budgets of projects that depended on it. By October 1989, Hubble's development costs stood at $1.6 billion, and millions more were being spent while Hubble waited for a ride to space.[6] A year later, with accumulated operations, servicing, and storage costs added in, the total costs had risen to $2.19 billion.[7]

Although demands on the budget were rising, available funding was not. NASA's budget was being squeezed on the other side by Congress, which kept appropriations nearly flat (adjusting for inflation) at $9 billion throughout the 1980s (see Fig. 6.2, p. 86). Even after launch, Hubble's relentless drain on the budget continued, as it was revealed that the telescope's

---

6. The Shuttle fleet was grounded in 1986 due to the Challenger accident. When flights resumed, a backlog of military and commercial payloads was given priority over Hubble. See Robert W. Smith, with contributions by Paul A. Hanle, Robert Kargon, and Joseph N. Tatarewicz, *The Space Telescope: A Study of NASA, Science, Technology, and Politics* (Cambridge: Cambridge University Press, 1993). Smith notes that Hubble's development costs by October 1986 were $1.6 billion in real dollars (p. 371). For budget details, see *NASA Historical Data Book*, vol. V, chap. 4, table 4.5, and vol. VII, chap. 4, table 4.6; available at http://history.nasa.gov/SP-4012/vol5/vol_v_ch_4.pdf and http://history.nasa.gov/SP-4012v7ch4.pdf (accessed 30 August 2016).

7. See D. Radzanowski, *NASA Under Scrutiny: The Shuttle and Hubble*, Congressional Research Service Report CRS-1990-SPR-0030, 13 September 1990, p. 8; values reported in real dollars.

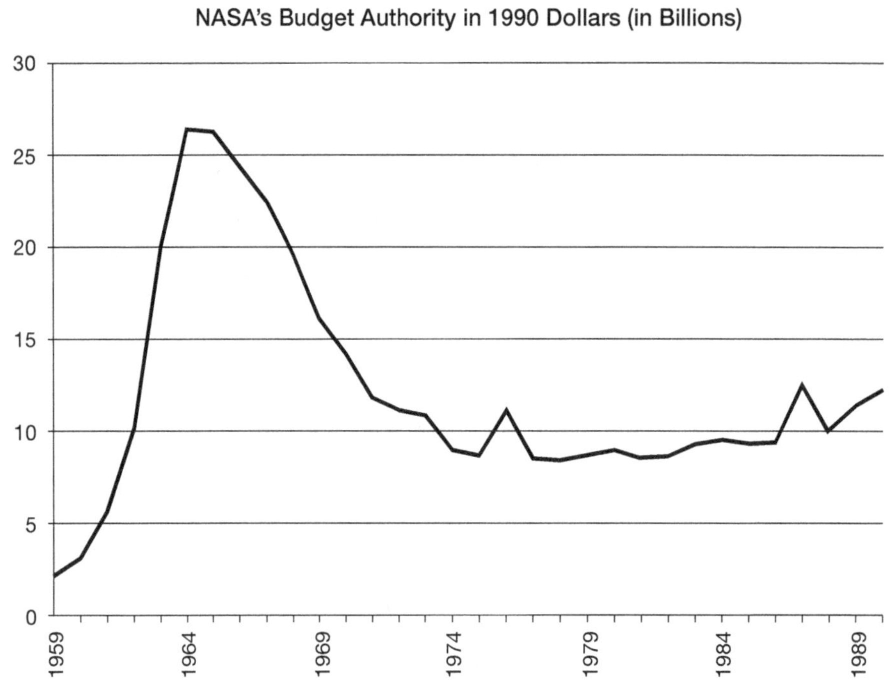

FIGURE 6.2. NASA's budget authority (billions, in 1990 dollars).

mirror was out of focus and would necessitate a Shuttle mission to repair it.[8]

Despite the widespread appeal of the Great Observatories concept, Congress had little appetite for more billion-dollar telescope projects until they could see the returns from their investment in Hubble. Even though AXAF had received a New Start for FY 1988 and the OMB had restored AXAF to the President's budget, Congress was reluctant to appropriate the funds to build it, perhaps in part because the company that had built Hubble's ill-fated mirror was also under contract to build AXAF's. To limit their exposure, Congress authorized only enough money for AXAF's mirror to be built. This was a test: If the AXAF team could demonstrate the critical technology by 1991, then further funding would be likely; if not, Congress would kill the project.[9]

The Great Observatories concept had successfully linked four distinct projects, in the hope that it would make them all easier to sell; indeed, all but SIRTF had been approved by 1989. But this approval also linked them with the "Hubble syndrome"—the potential for programs to consume resources in excess of those originally planned.[10] Hubble, with an initial price tag of $490 million, had ended up costing nearly $2.6 billion. The

---

8. Throughout the 1990s, Hubble's budget exceeded $150 million per year in real dollars for ongoing operations and three servicing missions (NASA's Office of the Inspector General, *Final Report on HST Cost Saving Initiatives*, Report IG-99-013, 19 March 1999, p. 6; available at *http://www.hq.nasa.gov/office/hqlibrary/documents/o43731805.pdf* (accessed 30 August 2016).

9. Wallace Tucker and Karen Tucker, *Revealing the Universe: The Making of the Chandra X-ray Observatory* (Cambridge, MA: Harvard University Press, 2001).

10. SIRTF SWG Meeting Minutes, 5 March 1990.

presumption was that if one Great Observatory had technical issues, cost overruns, or management problems, then the others were likely to as well. AXAF was estimated to cost $1.2 billion; SIRTF between $1.2 and $1.7 billion.[11] Until Hubble was fixed and AXAF passed its technical test, Headquarters could not go before Congress and credibly argue that SIRTF should be supported. As the last of the Great Observatories, SIRTF could not move forward until AXAF did.

Taking the Axe to AXAF

All of these billion-dollar telescopes had been conceived in more optimistic times. Now they were Charlie Pellerin's responsibility. As NASA's Director of Astrophysics, Pellerin reported to Lennard Fisk, the Associate Administrator in charge of OSSA.[12] Pellerin recalls, "Fisk called me into his office one day, and he showed me this chart. And he showed straight-line, 7-percent growth for space science. He said, 'This is the best we can hope for.'… The problem I had was that AXAF and Cassini [a Saturn mission] were peaking their funding.… They were peaking at the same time, and they were sticking up above that curve. Fisk said, 'What are we going to do about this?' I said, 'Thank you for sharing' and I went back to my office." Fisk was not going to cut other programs to stay within budget; the cuts had to come from AXAF or Cassini.[13] Pellerin was reluctant to take on Cassini—it had the support of the planetary community, which was concentrated at Caltech and JPL (where most of NASA's planetary missions were managed) and the Planetary Society (which had been founded by prominent scientists Carl Sagan of Cornell and Bruce Murray of Caltech). "I thought about it, and rightly or wrongly I decided that Cassini was much more powerful than me," Pellerin says. "At this time, AXAF was in Huntsville, at Marshall. The astrophysics program was all over NASA, in pieces. Marshall had AXAF, it had the Shuttle main engines, it had the Shuttle external tank. It had a space station coming up."[14] AXAF was just one of many projects competing for the attention of Marshall and the astrophysics community. Cassini, on the other hand, had the concentrated attention of JPL, Caltech, and the Planetary Society, each influential in its own right in the scientific community and in Congress. In short, Cassini had "[a]ll these political tools," Pellerin said, "that I didn't have [for AXAF]." The only tool he had was a fist to pound on tables at Marshall and demand that the project team find ways to reduce AXAF's $3 billion cost. "One day I got $200 million out of it," Pellerin remembers, "but it was still growing.… Here was the dilemma. The more it grew, the more it slipped, the more it slipped, the more it cost.… If [cost] becomes the issue, we're going to lose, and someone is going to cancel AXAF, and then the Great Observatories are gone, and SIRTF loses its physics rationale, and all that. So I had to find a way [to prove the project's value]."[15]

AXAF, Hubble, and SIRTF were all originally designed for a low-Earth orbit. While that orbit is relatively easy and inexpensive to reach, the usable observing time, or "efficiency," is poor, because the Moon, the Sun, or Earth is often

---

11. Presentation to Lennard Fisk, 28 June 1989.
12. Fisk was Associate Administrator for the Office of Space Science and Applications at NASA May 1987–June 1993; see Fisk interview, 8 September 2010.
13. Cassini did cut its costs by reducing the pointing capabilities of the planetary orbiter and eliminating CRAF, the comet mission. Details on the mission can be found in J. P. Lebreton and D. L. Matson, "The Huygens Probe: Science, Payload and Mission Overview," *Space Science Reviews* 104, nos. 1–4 (2002): 59–100.
14. Pellerin, interview, 19 March 2009.
15. Pellerin interview, 19 March 2009.

in the way. Pellerin realized that if he could get AXAF into a high-Earth orbit, then he would get a threefold improvement in efficiency. Even if he eliminated the Shuttle servicing and some of the more complex engineering components, which would diminish the facility's lifetime and overall capabilities, the improvement in efficiency would make up the difference. By lopping off major portions of AXAF, costs would come down; yet it could still deliver on the goals in the original project proposal. "I'm back to where I started," Pellerin said. "I've got the same science." Pellerin latched onto the idea of splitting AXAF into two missions, a proposal that the development team vehemently resisted.[16] The idea won out, however, and eventually a substantially lighter and cheaper AXAF mission, renamed Chandra X-ray Observatory, launched in 1999.

Resizing SIRTF

Meanwhile, SIRTF was experiencing mission creep. To secure the support of the planetary community, Cruikshank was recommending the addition of another instrument—a high-resolution spectrometer, which could be used to map the chemical composition of planets and other solar system objects. Frank Low was pushing for a high-resolution sky survey that would provide a more detailed sky map than IRAS had provided. Yet SIRTF, too, would have to be scaled back.

As with AXAF, Headquarters put pressure on the SIRTF team to find ways to shrink costs. One idea was to use a common spacecraft design for AXAF and SIRTF, although this proposal ultimately proved unworkable. Another scheme was to share costs with an international partner, as had been done in the highly successful IRAS project, the Dutch-U.S.-British infrared survey telescope launched in 1983. SIRTF SWG members met with infrared scientists around the world to explore ways they might collaborate. This was done at the urging of Headquarters, which was being pressured by Congress to leverage international participation and resources in lean economic times. Dutifully, the SWG had gone on what they called the "ISO death march," a four-day, four-country visit by several SWG members in spring 1988 to explore how the United States might participate in the Infrared Space Observatory (ISO), a European Space Agency project.[17] The SWG also met with a team of Japanese scientists, who had ideas for an instrument (and resources) they might add to SIRTF.[18] In the end, these initiatives did not pan out.

A more promising (and proven) tactic for reducing costs was to assign the best management team to the project. The SWG was responsible for figuring out what SIRTF needed to become a useful scientific instrument. But every brain also needs a body. The body of SIRTF comprised dozens of systems—including cryogenic, electrical, flight navigation, and data acquisition—that had to be engineered and then assembled into one spacecraft (see Fig. 6.3). This process was the responsibility of the project manager and the Ames project office. Later, additional teams would manage the launch, operate the spacecraft, and process the telescope data. There were thousands of parts and hundreds of people. It is safe

---

16. Pellerin interview, 19 March 2009; see also Tucker and Tucker, *Revealing the Universe*. AXAF was split into two missions in mid-1992, AXAF-I (imaging) and AXAF-S (spectroscopy). AXAF-S was canceled in 1993.

17. A handful of U.S. scientists were involved in ISO, including Martin Harwit, who was funded by NASA (SIRTF SWG Meeting Minutes, 11–13 May 1988).

18. The Japanese scientists were looking for a home for their instrument after the IRTS project was canceled in Japan; see Satio Hayakawa and Mamoru Saito, "Astronomy in Japan," *Astrophysics and Space Science* 99, nos. 1–2 (1984): 393–402. Eventually, the IRTS project was reinstated by the Japanese government, and a version of this instrument was flown, with participation by SIRTF/Ames scientist Tom Roellig.

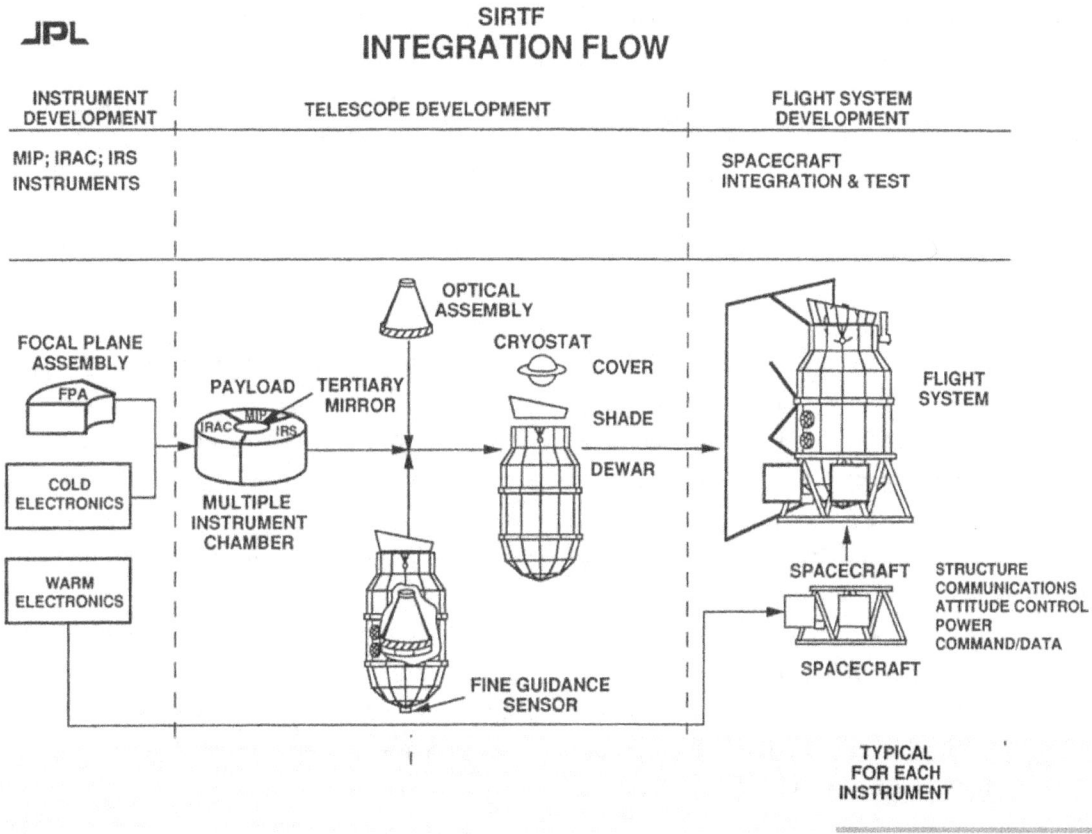

FIGURE 6.3. Integration flow for SIRTF systems (SIRTF Briefing for OSSA, 22 March 1990).

to assume that there will be problems in a complex project with untested technology and that some of these problems will be impossible to plan for. Therefore, it is vital that the project manager have broad experience and that the people around him have deep and diverse expertise to handle any problems.

Ames had never managed a $2 billion project. Marshall, JPL, and Goddard had. Ames employed 1,000 people. Marshall, JPL, and Goddard were all many times that size. Ames's only major space project was Pioneer, a series of planetary missions the last of which had launched in 1978. Did Ames currently have the necessary skills and staff to manage a project the scale of SIRTF? This was the question that several people at Headquarters were asking in the spring of 1989.

Losing Faith in Ames

SIRTF was the responsibility of the Space Science Division at Ames. Unfortunately for SIRTF, space science was not one of Ames's core elements. At Ames, it was aeronautics research and Shuttle-related programs, not space science, that received priority, as these were seen to be more in line with Ames's overall mission. Ames's management of Pioneer is a case in point. The Pioneer missions had generated broad popular interest, especially Pioneer 10, with its gold-anodized aluminum plaque depicting a man and woman, intended to show any extraterrestrial intelligence it encountered who the creatures were that had built this probe. However, despite public and scientific interest and substantial financial support from NASA, Ames's leadership provided only lukewarm support for Pioneer.[19]

The Ames Space Science Division had developed a strategic plan in 1985 to address this failing. Specifically, its leadership sought to "increase…[the division's] visibility and credibility both internally and externally" in order to acquire the funding and manpower to support their desired projects and to "develop a mindset" within Ames and Headquarters to "change the Division's status at Ames from one of 'benign neglect' or 'toleration' to being an important Center component."[20] By 1989, the division had not yet succeeded in changing the perceptions of Ames's management. The Director of Ames, William Ballhaus, provided Headquarters with a list of his Center priorities for 1989. "I think SIRTF was number 14 on the list," recalls Larry Caroff, who left Ames in 1988 to head the Infrared, Submillimeter, and Radio Astrophysics Branch at Headquarters. "Charlie [Pellerin] hit the ceiling about that and said 'Ames doesn't really want this, they don't care about it, I don't think they can handle it.'"[21] Larry Manning, who had worked on SIRTF and with Caroff at Ames before joining him at Headquarters, elaborates:

> The two Center directors that were involved—Ballhaus up until [February] 1988 and then [Dale] Compton—clearly saw Ames as a research center, and neither of them were project people. Dale's only project experience had been on IRAS, which had not been, from his viewpoint, a positive experience. I think he clearly wanted some indication from Charlie that Ames is going to get this, and Charlie wanted indication from Ames that Ames is going to be linked with the people that were needed to do it. They sort of butted heads.[22]

Pellerin was not worried about the ability of the SWG and the Ames project office to develop a good design. SIRTF's detector-development program was among the most advanced in the world for long-infrared wavelengths, and the scientists there had conducted dozens of mission options studies. Designing SIRTF was not what Pellerin was worried about—it was flying SIRTF. Pellerin was concerned that Ames did not have the staff to rapidly and adequately troubleshoot problems or the processes for averting problems with the contractors who were building SIRTF. It wasn't clear that Ames had enough people to do the job, and even if the Center did, it was undermined by the fact that the managers at Ames would not aggressively commit to SIRTF. According to Pellerin:

> Flight projects are a whole different deal. Flight projects are just antithetical to research. Researchers…are going to do the best of everything, and they're not going to use processes.… Flight projects have very rigorous processes for testing,

---

19. Witteborn interview, 2 September 2008.
20. Space Science Division Strategic Plans, 1980s, AFS1070.8A, Archives Reference Collection, FC5:D4, p. 5, NASA Ames History Office, NASA Ames Research Center, Moffett Field, CA.
21. Larry A. Manning and Lawrence J. Caroff, interview, Mountain View, CA, 3 September 2008.
22. Manning and Caroff interview, 3 September 2008. According to http://www.nasa.gov/offices/nac/members/Ballhaus-bio.html, Ballhaus served as Director of Ames (1984–1989) and also served as Acting Associate Administrator for Aeronautics and Space Technology at NASA Headquarters (1988–1989). At Ames, he was a research scientist (1971–1978), the Director of Astronautics (1980–1984), and chief of the Applied Computational Aerodynamics Branch (1979). Ballhaus worked for Lockheed Martin from 1989 to 2000, when he became the president of Aerospace Corporation in September of that year and then CEO in May 2001. A list of Center Directors that includes dates of their service is available at http://www.hq.nasa.gov/office/pao/History/director.html (accessed 30 August 2016).

for what they call validation and verification.... "Verification" means you meet the requirements and "validation" means that it's going to operate as intended and give the results.... It's not unusual for a space system to spend 18 months in test before it launches. Thermovac chambers that simulate space and all that stuff. So the project centers know how to do this and have the mindset to do it. Ames doesn't. They have none of the infrastructure for it. At Goddard or Marshall, they've got at least 1,000, maybe up to 2,000, engineers who are in what's called the functional organization.... And they're grouped by disciplines. And they're all experts in cryogenics or system engineering or mechanical systems. I think mechanical systems at Goddard has 800 people in that division. So when I had a contract that got in trouble—the same with JPL—those people go fix it. They go out to the contractor plant, and they bring in technical expertise for overseeing what the contractor is doing and then tell the contractor how to do it.... When Ames is in trouble, I have nobody to go to. I have no engineering staff. I have no one who's worked on a project since Pioneer, which last launched in 1978. I've got nobody there. So why do I want to expose myself to that risk?[23]

Pellerin wanted to transfer SIRTF to another Center. He had just finished the last of his battles with contractors over the construction of Hubble. (Hubble would launch in a year, and new battles would ensue over the imperfect mirror.) There had been problems with Lockheed, the contractor responsible for Hubble's system integration and test. This was due to incompatibilities between Lockheed and Goddard in their organizational cultures and testing processes.[24] An army of engineers from Goddard went to Sunnyvale and stayed until the issues were resolved. It was quite likely that SIRTF would end up at Lockheed, too, if for no other reason than that Ames was right next door to the Lockheed facility. "I was scared to death that Ames might end up doing SIRTF with Sunnyvale with nobody to watch," Pellerin said. "How am I ever going to get the oversight I need and the management I need?"[25] Although the contract for Hubble was huge by NASA's standards, it was small by Lockheed's. Goddard found it hard to get the best people and the attention of management at Lockheed.[26] The majority of Lockheed's revenue came from classified work. Pellerin was worried that Ames, given its size, would have even less influence over the contractor.[27]

---

23. Pellerin interview, 19 March 2009. Compare Pellerin's remark about JPL in this passage with the following entry in one of the Caroff notebooks: "JPL: center positioned as planetary, but also long wavelength astrophysics; would be enthusiastic to work with Ames on IR; lots of personnel (5300) and can 'store' about 1000 people by doing DoD work when a NASA project is not available." Note that this is a unique and flexible advantage for a Center working on projects that drag out at NASA (entry dated 9 November 1988, "IR/SubMM/Radio Branch, Book II, July 29, 1988, to November 3, 1989," Lawrence J. Caroff Notebooks, PP08.18, Box 1, Folder 3, NASA Ames History Office, ARC, Moffett Field, CA).

24. Smith, *The Space Telescope*, p. 365.

25. Pellerin interview, 19 March 2009.

26. Lockheed certainly has talented people—this is the firm that, in response to Sputnik, developed the first U.S. spy satellite at Sunnyvale, while the Burbank facility is where the U-2 and A-12/SR-71 Blackbird originated and gave us the term *skunk works* to designate a place of innovation; available at http://www.lockheedmartin.com/us/news/press-releases/2010/august/FirstSuccessfulCoronaRemo.html (accessed 30 August 2016).

27. Pellerin interview, 19 March 2009.

A Noncompetitive Competition

On 5 April 1989, a letter went out to the Directors of all of NASA's Centers announcing that a "non-competitive competition" would be held to determine which Center would be given management of SIRTF. The letter was from Pellerin's boss, Lennard Fisk, who indicated that Ames was the default choice, but there were caveats. Fisk wrote:

> I have decided that Ames, in consideration of its extensive and excellent study activities in support of SIRTF, will be given first consideration in this selection. For this reason, the Management Plans submitted by other centers (which are generally more experienced in this class of mission) will be used first as aids in evaluating the reasonableness of the Ames plan. If, after such evaluation, it is decided that the Ames plan is sound, realistic, and consistent with agency polices and plans, Ames will be selected as the management center. If, on the other hand, it is decided that Ames should not be selected, we shall proceed, on the basis of the material in hand, to select one of the other centers. It is planned that this final decision will be made by June 1, 1989, two weeks after receipt of your submissions.[28]

The decision, in part, had already been made. In Pellerin's mind, Goddard and JPL were the obvious choices. Both had been expressing their interest in the project since November 1988.[29] Goddard had developed the Einstein Observatory, the first X-ray space telescope launched in 1978, and was currently overseeing Hubble (in coordination with Marshall Space Flight Center, where AXAF was located). In addition, Goddard was already building IRAC, SIRTF's short-wavelength infrared camera, for Giovanni Fazio, the instrument's principal investigator. The other leading candidate was JPL, which wanted to be the home for infrared astrophysics (together with Caltech, which manages JPL for NASA). "JPL really wanted it badly," Pellerin said. "They romanced me hard."[30] Pellerin wanted to go ahead and transfer SIRTF to JPL.

People at Ames were upset by this decision—they had nurtured SIRTF for 18 years—and did not want to see it taken away just when it looked as if it would become a reality. Pellerin told the SIRTF team at Ames, "You don't have the depth of expertise I want when [the contractors] get in trouble.... It's not you guys. I don't have any difficulty with you.... This center doesn't have the depth to do it."[31] The Ames team thought they could handle it and demanded a competition. Though Pellerin was doubtful that Ames could win, he agreed to it and Fisk oversaw the process.

Four Centers bid for SIRTF—Ames, JPL, Goddard, and Marshall—and submitted a project management plan. Each expected to handle the project at their own Center, but Marshall's plan proposed teaming with Ames to manage the SIRTF Science Center, which would handle science program selection, scheduling, data acquisition, and processing after launch. In the little time that everyone had to put together plans,

---

28. Lennard Fisk to NASA Center Directors, dated 5 April 1989, in Science Working Group minutes and presentations of 10–12 July 1989, pp. 88–90.

29. Lawrence J. Caroff Notebooks, entries dated 9 November and 22 November 1988, PP08.18, Box 1, Folder 3, "IR/SubMM/Radio Branch, Book II July 29, 1988 to November 3, 1989," NASA Ames History Office, NASA Ames Research Center, Moffett Field, CA.

30. Pellerin interview, 19 March 2009.

31. Pellerin interview, 19 March 2009.

Marshall had not coordinated this with Ames, so it was difficult to evaluate this teaming arrangement. Pellerin would have resisted it anyway. "Anytime someone says, 'I'm going to set up an organization to get the best people from everywhere,' you run away," he said. "I'm going to set up an organization with clean interfaces, and they're going to find the people.... Inter-Center interfaces are disastrous.... Nobody wants to be subordinate.... With a Center and a contractor, it's no problem. The Center is in charge, the contractor is the second-tier party.... But when it's even/even, and neither one wants to be bossed by the other, nobody's in charge."[32] Pellerin insisted that one Center get SIRTF, all of it.

At the end of June, representatives of the four Centers were invited to present their plans to Fisk and Robert Rosen, the Associate Administrator for NASA's Office of Aeronautics and Space Technology. The criteria were, in equal measure, the Center's commitment to SIRTF, its scientific/technical understanding and past performance in building infrared instruments, the qualifications of the project team at that Center, and the adequacy of its management plan. "Adequacy" included considerations of the Center's past performance in managing billion-dollar projects, and a projected cost and development schedule in line with other Centers'. In addition, the plan needed to specify which portions of the program were to be outsourced to contractors and how their performance would be monitored, what the interfaces would be between organizations, how the systems would be integrated, and where the expertise and personnel would come from to troubleshoot or manage crises that might arise.[33] Watching this process unfold, Larry Caroff was dismayed that his former colleagues at Ames might lose the project to which they had dedicated so much of their careers. Caroff was joined in his concern by Fred Gillett, who was wrapping up two years at Headquarters as a Visiting Senior Scientist for the Infrared and Radio Astrophysics Branch (a branch he helped establish). Both Caroff and Gillett understood the concerns of their current colleagues at Headquarters and so sought to provide insight into this process. Although projects had occasionally been transferred from one Center to another, there had never been a competition like this. The rules had to be invented. Gillett wrote to the Ames team that

> I and others are trying to make sure that this process is as fair as possible, but you should recognize at the start that there is going to be a tremendous variation in the parameters that describe the definition and development phases.... That means that there will be a tremendous latitude for interpreting the reality of proposed resource plans. Therefore, Ames's proposal is going to have to do more than just be consistent with what is adjudged as appropriate...because "appropriate" is in large part going to be in the eye of the beholder. The way to win this non-competitive competition is to make the proposal so compelling that it would seem foolish not to support it.[34]

The Ames proposal had to make Pellerin look smart in recommending them, and it had to dispel the notion that Ames management was ambivalent about the SIRTF program.

Ames management made their strong support very clear. Nevertheless, serious reservations

---

32. Pellerin interview, 19 March 2009.
33. Fisk, 5 April 1989, letter.
34. E-mail from Fred Gillett to D. Compton, J. Sharp, and P. Dyal, "SIRTF Management Decision" (message # LJIJ-2807-7450), dated 14 April 1989.

TABLE 6.1. Criteria of Center competition with scores ranging from 1 (low) to 5 (high).

Criteria	Ames	Goddard	JPL
Center Commitment • Priority of SIRTF and Center commitment • SIRTF project access to top levels of management • First-rate team/program phaseout to accommodate SIRTF	4	3	4
Scientific/Technical Expertise • Demonstrated scientific/technical understanding of SIRTF program • Scientific/technical approach planned for SIRTF • Scientific/technical Center experience and past performance	5	5	4
Management Plan • Management Center experience and past performance • Program schedule and resource requirements • SIRTF project organization and management approach • Assignment of responsibilities (in house or contract) • Methods of project monitoring and control (in house or contract) • Technical problem solving and crisis management • The role of other Centers • Management, scientific, and reporting interfaces between project/program • Cost and availability of proposed facilities	1	5	5
Project Team • SIRTF definition phase project team (phase B) • SIRTF development phase project team (phase C/D) • Additional personnel proposed for SIRTF project team	3	4	4

remained about whether Ames had the necessary resources to solve technical problems and effectively handle crises. Marshall's project plan did not provide a sufficiently detailed account of how a teaming arrangement with Ames would work, and the reviewers found it difficult to score their plan. JPL and Goddard had similar strengths, but ultimately JPL demonstrated greater interest in managing SIRTF. An average score for all criteria is presented in Table 6.1.

It is likely that the competition was over before it began. Ames lost and JPL was given responsibility for taking SIRTF through to the next stage. In some ways, it was not unlike Pellerin's experience with the straw vote on New Starts. The cards had already been dealt, and all that was left was to play the hand. Regardless of whether the competition was fair to Ames or slanted against it, the process provided a rare and valuable opportunity for Headquarters to bring the expertise of four of its major project Centers to bear on SIRTF and take stock of the program. By comparing the different cost estimates and project management approaches, executives in Fisk's and Pellerin's OSSA could triangulate the likely costs and possible problems of SIRTF.

Seismic Shift

At the end of September 1989, Headquarters began the formal transfer of SIRTF from Ames to JPL.[35] The project experienced a physical dislocation of 300 miles, roughly paralleling the San Andreas fault line, from Mountain View to Pasadena, California. After the rumbling stopped, the project found itself relatively intact.

---

35. Lawrence J. Caroff Notebooks, entry dated 9/26/89, PP08.18, Box 1, Folder 3, "IR/SubMM/Radio Branch, Book II July 29, 1988 to November 3, 1989," NASA Ames History Office, NASA Ames Research Center, Moffett Field, CA.

TABLE 6.2. Full mission life cycle.

Pre-Phase A	Conceptual study	
Phase A	Preliminary analysis	
Phase B	Definition	• System requirements review • System design review • Non-advocate review
Phase C/D	Design and development	• Preliminary design review • Critical design review • Test readiness review • Flight readiness review
Phase E	Operations phase MO&DA (Mission Ops & Data Analysis)	• Primary mission • Extended mission

Prior to the competition, JPL had begun working with Ames to help define a data communications plan under various orbit scenarios. As a result, JPL scientists were already familiar with SIRTF's design when the laboratory was awarded the management contract. While Ames personnel would rather have retained the management of SIRTF, they wanted SIRTF to succeed and helped to make the transfer as smooth as possible.

JPL wanted Mike Werner to remain as project scientist. Werner wanted to, as well; for two years he commuted to JPL from Northern California, where his son was finishing high school. The only other person to transfer from Ames to JPL was Peter Eisenhardt, who was a co-investigator on the IRAC instrument. Werner had hired him after Eisenhardt completed his Ph.D. at the University of Arizona in 1984.

The bulk of the knowledge about SIRTF resided in the Science Working Group.[36] This was both unusual and fortunate, as it really didn't matter to the SWG which Center was responsible for the project. The SWG's scientists were at the cutting edge of the infrared detector technology upon which SIRTF's instruments relied. The flight Centers—JPL, Goddard, and Marshall—each had relevant skills for designing and operating the spacecraft on which the SWG's instruments would be installed.

Another reason that the move to JPL was not more disruptive is that SIRTF's design was not that mature. Despite the fact that SIRTF had been under development for almost two decades, it was still in Phase A. Since 1971, all of the work on SIRTF had been to develop a concept and conduct preliminary analyses. SIRTF was just on the cusp of moving into Phase B, which is NASA nomenclature for the engineering-design portion of the project. After successfully completing Phase B, NASA might then authorize a New Start and enter Phase C/D to build and launch the facility (see Table 6.2 for details). Up to this point, the SIRTF SWG and project office members had mostly worked on feasibility studies, tradeoff analyses, and detector development. These activities were useful for developing technical knowledge but did little to move the project forward. Indeed, there was a sense that some were simply "bring me a rock" exercises, designed to keep the team together and busy while Headquarters

---

36. Michael W. Werner, interview by author, Pasadena, CA, 15 December 2008; and Caroff and Manning interview, 3 September 2008.

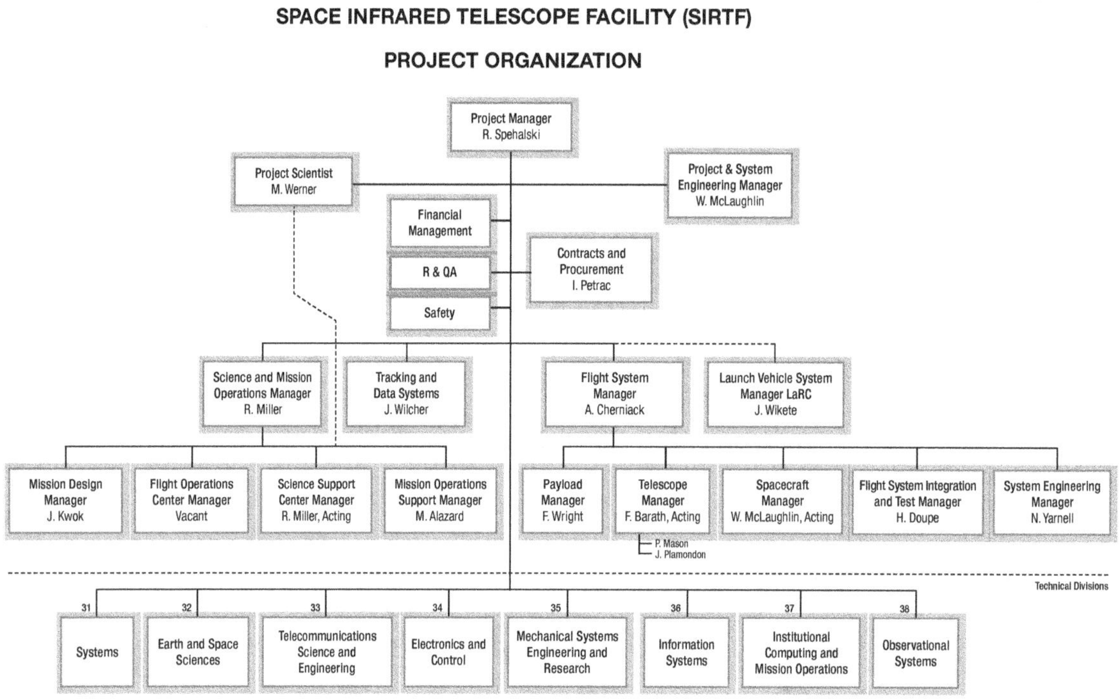

SPACE INFRARED TELESCOPE FACILITY (SIRTF)
PROJECT ORGANIZATION

```
                                    Project Manager
                                    R. Spehalski
                                         │
        ┌────────────────────────────────┼────────────────────────────────┐
  Project Scientist                      │                         Project & System
  M. Werner                              │                         Engineering Manager
                                   Financial                       W. McLaughlin
                                   Management
                                         │
                                   R & QA ── Contracts and
                                             Procurement
                                             I. Petrac
                                   Safety
                                         │
   ┌────────────────────┬────────────────┼────────────────┬──────────────────┐
  Science and Mission   Tracking and    Flight System    Launch Vehicle System
  Operations Manager    Data Systems    Manager          Manager LaRC
  R. Miller             J. Wilcher      A. Cherniack     J. Wikete
```

| Mission Design Manager J. Kwok | Flight Operations Center Manager Vacant | Science Support Center Manager R. Miller, Acting | Mission Operations Support Manager M. Alazard | Payload Manager F. Wright | Telescope Manager F. Barath, Acting | Spacecraft Manager W. McLaughlin, Acting | Flight System Integration and Test Manager H. Doupe | System Engineering Manager N. Yarnell |

P. Mason
J. Plamondon

Technical Divisions

| 31 Systems | 32 Earth and Space Sciences | 33 Telecommunications Science and Engineering | 34 Electronics and Control | 35 Mechanical Systems Engineering and Research | 36 Information Systems | 37 Institutional Computing and Mission Operations | 38 Observational Systems |

FIGURE 6.4. Organization chart of SIRTF (c. 1990) (SIRTF Briefing for OSSA, 22 March 1990).

worked to obtain the necessary financial and political support for a New Start.[37]

One of the first tasks after the project arrived at JPL was to prepare the request for proposals (RFP) for Phase B. The RFP was a solicitation for contractors to turn the Phase A conceptual design into an engineering design. This activity had started at Ames and was continued at JPL as a way to show that the project hadn't lost momentum as a result of the move. The RFP was issued with the permission of NASA Headquarters on 1 June 1990.

As the excitement over SIRTF wore off a little, JPL managers realized that they first needed to have their own internal review before bringing in outside contractors. JPL had assigned a seasoned project manager to SIRTF, Dick Spehalski, who had previously managed several billion-dollar missions. In October 1989, Spehalski convened an internal review of SIRTF's engineering and systems integration issues, drawing JPL representatives from all the major engineering areas (see organization chart in Fig. 6.4).[38] This joint review provided an opportunity to consider SIRTF in a more reflective way than had been possible during the brief management competition.

As Spehalski was bringing together the JPL team to assess SIRTF, the Agency's overall priorities and performance were being reviewed at the behest of Vice President Dan Quayle by a 12-member committee chaired by Norman Augustine. In December 1990, they presented their findings in the *Report of the Advisory Committee on the Future of the U.S. Space Program.* As noted earlier, the Augustine commission gave NASA a mixed rating, affirming space science as one of NASA's core elements but criticizing the Agency for its lack of a coherent national space

---

37. McCreight interview, 2 September 2008.

38. Dick Spehalski, SIRTF JPL Internal Review, October 1989.

policy. The committee was of the opinion that "NASA is oversubscribed in terms of the projects it is pursuing, given its financial and personnel resources and the time allotted to pursue them."[39] In this climate, which was also chilled by the possibility of war in the Persian Gulf, Congress asked NASA to defer Phase B activities for SIRTF. Once again, SIRTF had to hit the brakes.

Attention to Advocacy

To successfully sell a billion-dollar project, one needs both political support and financial resources. At the end of 1990, SIRTF didn't have enough of either one. The SWG knew that it had to reduce SIRTF's costs (a fact that Pellerin would regularly remind them of). Certainly, SIRTF's costs had to align with the available funding, but how much was available depended on the mood in Congress and at Headquarters. So, with one eye on the budget, the SWG began to focus on managing the mood of stakeholders through deliberate advocacy.

The SWG had learned the hard way about not having explicit support in the Field report. With a new decade, there would be a new survey. The 1990s decadal survey was chaired by John Bahcall, as enthusiastic a supporter of space-based telescopes as was his older colleague Lyman Spitzer. The SWG took an aggressive stance to ensure that the Bahcall report gave SIRTF sufficient support; the goal was not merely for SIRTF to be a priority, but for it to be the *top* priority.

The SWG was aided by having Fred Gillett and Jim Houck, two major supporters of SIRTF, as chair and vice-chair, respectively, of the Infrared Astronomy Panel of the 1990s survey committee. Gillett, as noted above, had helped to establish the Infrared branch at Headquarters and was a good friend of Frank Low, one of the SIRTF facility scientists. Low and Gillett had both worked on IRAS, where Low was responsible for the cryogenic technology and Gillett for the detectors. Gillett and Low were both brilliant but opposite in temperament. Low relied on his intuition, preferring that things be done his way, and (to the irritation of others) he was usually right. He was argumentative but deeply committed—playing the role of the "loyal opposition."[40] In contrast, Gillett was meticulous, polished, and very skilled at building consensus. Choosing Gillett as chair was a decision with which even Low couldn't disagree. Jim Houck was also a favorable choice for the infrared panel. Houck was a member of the SWG and PI for the spectroscopy instrument on SIRTF. Houck had even been a member of the 1970s decadal survey (Greenstein report). Back then, there were only a handful of infrared scientists. Although more established scientists ultimately set the agenda for the 1970s, the idea of a SIRTF-like project was not new. Houck had introduced it then but would now be in a better position to address objections.

Gillett and Houck succeeded in their mission.[41] Not only was SIRTF named as *the* top priority in the Bahcall report, but the 1990s were named the "Decade of the Infrared."[42] There had never been a clearer statement of scientific support for SIRTF. Now the SWG just needed to secure funding for the project.

While Congress was advising NASA to defer the Phase B activities for SIRTF, the SWG was

---

39. Augustine report, 1990, page 19.
40. Michael Werner, "Frank Low's Contributions to the History of Infrared Astronomy," presentation at the 214th Meeting of the American Astronomical Society, 7–11 June 2009, Pasadena, CA.
41. Another advocate was Charles Beichman, who had an indirect position of influence on the decadal review. Beichman, an infrared astronomer at Caltech/JPL and a key participant in SIRTF's IRAC instrument, moved to Princeton for a year to help John Bahcall manage the decadal review process.
42. Bahcall report, p. 75.

getting advice on how to lobby Congress. JPL's director for Earth and Space Science, Charles Elachi, and the SIRTF program manager at Headquarters, Art Fuchs, recommended to the SWG that it appoint someone to be the chief advocate for SIRTF.[43] Such a strategy had worked for AXAF; Harvey Tananbaum, of the Harvard-Smithsonian Center for Astrophysics, was a tireless and relentless advocate for AXAF and was free to promote the project because he was not a federal employee. It should be noted that, as government employees, NASA personnel are not permitted to lobby or directly persuade elected officials to support a particular program (Pellerin's own AXAF maneuvers notwithstanding). NASA's priorities are dictated by the White House and Congress, and government employees are meant to carry out those directives. However, not all members of the SIRTF SWG were employees of NASA or the government. The three Principal Investigators were all employed by universities: Jim Houck (Cornell), George Rieke (University of Arizona), and Giovanni Fazio (Harvard-Smithsonian Center for Astrophysics). As such, they were not prohibited from lobbying—nor, for that matter, are JPL employees, who are directly managed by Caltech, not NASA.[44] However, the three PIs were all busy serving on professional committees, teaching courses at their home institutions, and developing their instruments for SIRTF.

Therefore, starting with its April 1991 meeting, the SWG was joined by a new member, Marcia Rieke.[45] Like her husband George, Marcia Rieke was an accomplished astronomer, instrument builder, and faculty member at the University of Arizona. Both had earned their doctorates in Boston in the 1970s, George at Harvard, Marcia at MIT. They later met at the University of Arizona and eventually married. While George was busy developing detectors for SIRTF, Marcia was doing the same for Hubble. Marcia's detectors, among the first 128- by 128-pixel arrays in infrared astronomy, were incorporated in the Near Infrared Camera and Multi-Object Spectrometer (NICMOS) and installed on Hubble during a servicing mission in 1997. Hubble, primarily an optical telescope, could now see in the dark (at least in the near-infrared spectrum), thanks in part to Marcia.

Rieke was enthusiastic about her new role as SIRTF's chief advocate. At the earliest opportunity, she visited DC. Her timing could not have had been worse. The American Astronomical Society was holding its 177th meeting in Philadelphia from 13 to 17 January 1991. "I took the train down to Washington, DC," Rieke said. "I was going to talk with the folks at Headquarters about what approach we'd use in doing advocacy and presenting the SIRTF project to people on Capitol Hill so they'd understand it, and so on. Well, it was kind of a dead-on-arrival meeting, because I walked in the door and they said, 'There's a letter from Senator Mikulski [saying] that these projects are all canceled.'"[46] To top it off, that week, a U.S.-led coalition began the bombing attack on Iraq known as Desert Storm.

---

43. SIRTF SWG Meeting Minutes, 11–13 December 1990.

44. JPL is a Federally Funded Research and Development Center and is managed by Caltech under contract with NASA. Furthermore, Caltech is a private research university. Because of this arrangement, unique among the NASA Centers, employees of JPL/Caltech can lobby government officials on behalf of themselves or their organizations.

45. SIRTF SWG Meeting Minutes, 9–11 April 1991.

46. Barbara Mikulski, Democratic senator from Maryland who at the time of this writing is nearing retirement, was a member of the Appropriations Committee from 1987 to 2015. She is a strong advocate for NASA: Goddard is in her state, and Hubble alone has brought hundreds of jobs to her state. While the quotation reflects Mikulski's influence, she would not have had unilateral power to cancel a mission. Instead, cancellations result from legislative or report language of the Appropriations Committee.

Rieke had become SIRTF's advocate just as NASA Headquarters was told to cancel SIRTF. People avoided mentioning SIRTF in the halls of Headquarters.[47] Associate Administrator Len Fisk scraped together discretionary funding and funneled it to the "Infrared Astronomy Mission" that everyone formerly with SIRTF was now working on.[48]

Project Manager Merry-Go-Round

The "Infrared Astronomy Mission" was sustained by scraps from Headquarters' and JPL's discretionary funding. The detectors were the crown jewels of the telescope, and spending time and money on their development was important and useful work. But without a real project to hold the team together, JPL began to take seasoned managers from SIRTF and reassign them. For better or worse, JPL had plenty of work for experienced managers to do. The high-gain antenna on the Galileo probe failed to deploy in April 1991. Spehalski, who had been managing the Galileo project one way or another since the mid-1970s, was put back on to help troubleshoot Galileo, leaving SIRTF after only one year as its project manager.[49]

SIRTF was no longer a major project but simply a detector-development initiative. So JPL assigned James A. Evans to be the project manager of this "sort-of SIRTF"; his formal title was manager of the Astrophysics and Fundamental Physics Preprojects program. Evans's two decades of experience managing the development of tanks for the Army did not make him a particularly good fit for SIRTF or for infrared detector development in general, but he remained with sort-of SIRTF until October 1993, when he became assistant laboratory director of JPL's Office of Technology and Applications.[50]

Moving managers around was a disruptive but increasingly necessary practice. Media and congressional scrutiny were now a regular part of operations, especially when something went wrong with a project for which taxpayers were expected to pay a billion dollars or more. The best project managers were in high demand to quickly correct technical issues and avoid public-relations disasters. Two months earlier, in August 1993, mission operations engineers had lost contact with Mars Observer just as it was nearing the planet.[51] The public had all but forgotten about the probe during the year it took to reach Mars. Presumably, it exploded because of a fuel leak while attempting to achieve orbit. Before the explosion of Challenger and the Hubble mirror debacle, such a loss might have received little attention. Now it was front-page news.

There was no project more important to NASA in the autumn of 1993 than the repair of Hubble's mirror. Hubble had launched in 1990 with a mirror whose outer edge was too flat by 4 microns (roughly equal to one-twelfth the thickness of a human hair). NASA did not catch the problem before launch; the polishing process was a classified military secret, and NASA had relied on the contractor's test reports. The result of this spherical aberration was a telescope that

---

47. Marcia J. Rieke, interview by author, Long Beach, CA, 6 January 2009; also Werner interview, 15 December 2008.
48. The designation "Infrared Astronomy Mission" is used instead of "SIRTF" on the project Gantt charts (see, e.g., SIRTF SWG Meeting Minutes, 11–13 December 1992).
49. SIRTF SWG Meeting Minutes, 16–18 March 1992.
50. SIRTF SWG Meeting Minutes, 16–18 March 1992.
51. Mars Observer Mission Failure Investigation Board Report (T. P. Coffey, chairman), National Aeronautics and Space Administration, 31 December 1993; available at *http://spacese.spacegrant.org/Failure%20Reports/Mars_Observer_12_93_MIB.pdf* (accessed 30 August 2016).

could not focus properly. NASA called on experts at several Centers and on contractors to develop the solution, the Wide Field/Planetary Camera Number 2 (WFPC2), which functioned much like a pair of prescription glasses for correcting astigmatism. The servicing mission occurred in December 1993, during which Shuttle astronauts fitted Hubble with the new instrument, resulting in some of the most spectacular and sublime images of space anyone had ever seen.

The person responsible for the WFPC2 program at JPL was Larry Simmons, and a lot was riding on him. He had to coordinate the efforts of dozens of organizations and hundreds of people, who had 18 months to build WFPC2 and train the Shuttle astronauts to install it on Hubble, 350 miles above Earth—all under intense scrutiny from the press, the public, and Congress. Simmons explained to NASA Administrator Daniel S. Goldin how what they were doing was "going to save the Hubble." In response, Simmons remembers, Goldin, "nose to nose almost, looking me in the face, putting his glasses up on his forehead, [said,] 'What you're doing is going to save NASA.' I mean, he was adamant. So we had a lot of oversight."[52] Simmons would become SIRTF's next project manager.[53]

---

52. Larry Simmons, interview by author, Pasadena, CA, 18 February 2009.

53. In a personal communication dated 8 November 2010, Michael Werner commented:

> WF/PC-2 was under development at JPL when the spherical aberration problem was discovered. The camera was redesigned during the fabrication process to correct the images.... When Simmons came to work on Spitzer after the success of WF/PC-2, he brought with him several other excellent engineers—notably Steve Macenka and Jim Fanson (and eventually Dave Gallagher)—who had contributed to the success of WF/PC-2 and were really, really capable. I learned from this that one of the attributes of a good Project Manager is that he attracts good people to work with him, who may follow him from project to project.

CHAPTER 7

From Orphan to Poster Child

When Larry Simmons came on board in November 1993, the SIRTF SWG was doing what it could to make the pendulum swing back its way. Nearly three years had passed since Marcia Rieke's first visit to NASA Headquarters in January 1991; on subsequent visits, she found the Agency's administrators to be somewhat more receptive. The Gulf War was now over, and Congress was more willing to hear about new space-related initiatives, but on the Hill, NASA was perceived as inefficient, slow, and bloated. NASA—the innovative darling of the 1960s—had become, in the worst sense of the word, a bureaucracy.

To address this perception, the White House had appointed a new NASA administrator, Dan Goldin, to cut costs and reduce risks across the entire Agency. Goldin replaced Richard Truly, an admiral and former astronaut, who had been brought in to start the cost-cutting process by first taming the problems with the Space Shuttle and Hubble programs. Goldin took the reins on 1 April 1992. He had had 25 years of experience in the defense industry, most recently as the vice president and general manager of the TRW Space and Technology Group. He brought to NASA such management concepts as Total Quality Management and a greater reliance on outsourcing. During his tenure (1992–2001), he cut staffing levels, directed more funding toward science missions, and instituted a new policy on projects. In the Goldin era, projects could no longer be so large that they consumed the majority of the Agency's budget or caused a public outcry when they failed. Instead of a few large projects, there would be many small ones, resulting in more opportunities to experiment with new technologies and in shorter development cycles.

Faster, Better, Cheaper

Goldin's project philosophy came to be known as "faster, better, cheaper."[1] SIRTF was certainly better. But it wasn't faster or cheaper. Even before Goldin's arrival, Charlie Pellerin and the SWG had been searching for ways to reduce the cost.

---

1. Goldin's approach is probably best viewed as beneficial in the short term but problematic in the long term. A number of scientists have criticized the "faster, better, cheaper" approach. The Columbia Accident Investigation Board Report (6 vols. [Washington DC: Government Printing Office, 2003], available at *http://www.nasa.gov/columbia/caib/html/start.html*) argued that this policy was too aggressive and had undermined safety and quality at NASA, citing it as a cause of the second Shuttle disaster. For an in-depth discussion of Goldin's policies, including their benefits, see Howard E. McCurdy, *Faster, Better, Cheaper: Low-Cost Innovation in the U.S. Space Program* (Baltimore: Johns Hopkins University Press, 2001).

Pellerin wanted nothing more than to complete the Great Observatories. Thus SIRTF needed to be ready with a plan when Congress was once more receptive.

SIRTF would probably never be faster. Although the project had changed, the acronym had been the same for two decades. Fisk suggested giving it a new name as a way to reinvigorate the project. Unfortunately, doing so would wipe out all the advocacy and name recognition that had been achieved. The name remained.

SIRTF could perhaps be cheaper. A month before Goldin came on board, the SWG had unveiled a slimmer SIRTF.[2] As with any cargo, its weight largely determines how much it will cost. The larger the telescope, the more complex and costly it is to build and the more powerful and costly the rocket needed for launch. At the cheaper end, IRAS cost $400 million and was put into a 900-kilometer orbit with a Delta rocket. By contrast, Hubble—which cost $2 billion, weighed 24,500 pounds (11,113 kilograms), and spanned 43.5 feet (13.3 meters)[3]—was so large that only the Shuttle could accommodate it. By 1989, when the SIRTF project arrived at JPL, launching on the Shuttle was no longer the favored option, so SIRTF was configured to launch with a Titan rocket, the standard choice for large payloads. But Titan rockets are also expensive. By 1992, SIRTF had gone on a diet and was small enough that it could now fit on an Atlas rocket, which brought down costs. But the price for SIRTF was still a hefty $1.2 billion. Even in its current configuration, it would never satisfy Goldin's criteria.

Going in Circles (Around the Sun)

Although shrinking SIRTF to fit on an Atlas rocket didn't entirely solve the budget problems, reducing the weight revived the possibility of launching SIRTF into an even higher orbit. The SWG had considered nearly a dozen orbit configurations by this point.[4] The perfect orbit was one that allowed for maximum unobstructed viewing and could be reached with a (cheap) launch vehicle. SIRTF was at that point designed to orbit Earth, with all the concomitant interference from the planet itself, plus its Sun and Moon. Earth orbit was easy to reach with most launch vehicles and provided plenty of opportunities to download data from the telescope as it passed over ground stations. From an observational perspective, however, it was inefficient.

Johnny Kwok had a radical idea. He had heard the scientists' argument that the farther the telescope could get from Earth, the better it would be for astronomical observations. This started Kwok thinking about a "trailing-Earth orbit," in which the telescope trails Earth around the Sun. This type of orbit had never been tried before. Kwok's insight was that, like two cars racing around a track, the telescope would follow behind Earth as they were both pulled along by the Sun's gravity. Although they would have the same orbit, the telescope would be a little slower than Earth. After six or seven years, the distance between them due to drift would likely be too great for communications; after 60 to 70 years, the gap would close as Earth lapped the telescope. Kwok, who wrote his doctoral dissertation on orbital mechanics and has worked at JPL on numerous missions, said:

---

2. Project-plan briefing presented by James Evans, SIRTF (JPL) Project Manager, to George Newton, NASA (Headquarters) Program Manager, June 1993.

3. Telescope specifications and total cost come from *http://hubblesite.org/the_telescope/hubble_essentials/quick_facts.php* (accessed 30 August 2016).

4. Rieke, interview, 9 June 2009.

It gave me an idea about how do I leave Earth. Well, from a simple, dynamical perspective, to leave Earth is very simple. You just give it enough energy, and with a simple equation that you can figure out, you just escape. But the problem with escaping is that, Well, which way do I escape to? That's something you had to model. Do I shoot the telescope in the Sun's direction, up or down?… In 1991, I was a supervisor in the mission-design section. I didn't have time to play around with modeling these things. One day, a summer student from Colorado came in, came to my office. His name was Cesar Ocampo—he's now a professor at the University of Texas, my old school…. He was a summer student in a different group and basically told his supervisor he didn't like the assignment he was given. The supervisor said, "Well, go talk to Johnny. He's always thinking about stuff that he may not have time to do himself." So Cesar came, and I talked to him. As an undergraduate he'd studied basically the same stuff that I'd studied, so it was very easy to talk to him and explain things to him. What came to my mind about SIRTF was this high-Earth orbit. The project was happy with the way it was, but I knew that a simple calculation says that high-Earth orbit was not a very good orbit to be in from an energy perspective. But if I let SIRTF escape, I can actually use a smaller rocket; I mean, that calculation is very simple. But the thing I didn't have time to figure out was, Which way do I send it [into solar orbit]?… [I]f the Sun [has] enough gravitational attraction, SIRTF may come back and hit the Earth…. [I told Cesar to model it], shoot it in different directions and sample all these variables, and then tell me what energy and direction you want to give to this spacecraft so it will escape [Earth], but escape at the minimum rate—because I didn't want it to leave Earth too fast, because then I couldn't talk to it. My data rate would go down if it got too far away from Earth. So the goal was to find the minimum rate at which it would leave Earth but without coming back and hitting Earth. Cesar knew exactly what I wanted, so over the summer he developed a model [and] produced a very nice report. In it, there was a plot that showed exactly what it took to do this, with the minimum distance. Then he left after the summer. I put that report away, because no one had asked for it. But I knew the answer.[5]

About a year later, Mike Werner was holding another SWG meeting and was looking for new ideas for SIRTF. James Evans, the current project manager, dropped by Kwok's office and asked him if he might have anything to contribute. Kwok said, "Yes, sure. I think you can change the orbit." When the agenda came out, Kwok's name was not on it. He figured the SWG was too busy or didn't want to hear about yet another orbit from a lower-level supervisor. It turned out that Kwok had been left off the agenda inadvertently. The SWG was interested, but the schedule was now full. "I ended up doing it during lunchtime," Kwok said. "I told them my… concept—sending a telescope away from Earth and just drifting in space." His presentation was met with silence. Frank Low was the first to speak, and Kwok braced himself. "Frank Low is another one of these brilliant astronomers, but he's very critical… he didn't have much respect for engineers…. What happened was that… Frank Low jumped up and said, 'That was the best idea I've heard all day.' So everybody starting rallying behind the [heliocentric, Earth-trailing] concept,

---

5. Johnny Kwok, interview with author (Session I), Pasadena, CA, 26 February 2009.

and then we went off and in the next few months we just redesigned the system so that it would come down in mass and in cost."[6]

Kwok's idea for a heliocentric, Earth-trailing orbit was very efficient and maximized the available observing time. SIRTF needed only to be made light enough to launch on an Atlas rocket. Even if some capabilities had to be cut to achieve the weight reductions, the improvement in efficiency might make up the difference. It gave the SWG some breathing room, despite the budgetary constraints.

The overall cost of the mission remained a problem. "Still we were stuck," Pellerin said.[7] A recurring question was whether SIRTF could be broken up into a few smaller missions (as had been done with AXAF), thereby making the whole easier to fund. The SWG duly conducted a "meiosis study" to see how SIRTF could be split up, but the cost savings were not compelling and the idea was dropped.[8]

However, the investment made over the preceding years in developing infrared detectors meant that they were now far more powerful than they had been when SIRTF was first conceived. Could the mission be reduced without losing any science? "It's the same argument as AXAF," Pellerin said. "Detector technology is a hundred times better [compared to 1983, when the SIRTF Announcement of Opportunity came out]. So I can make something one-hundredth the size and get you what we [originally contracted]. I don't want to have to do that. Just cut the cost by a factor of 3 or 4, and I'll be happy.... [The SWG] all pissed and moaned, just like the AXAF guys. They wanted the big thing. It took months and months, and they finally came around."[9]

By the time the SWG came around and accepted that NASA would never give them a billion dollars, it was November 1993, and many of SIRTF's supporters had left the stage. Bill Clinton was president, the leadership in the House and Senate had shifted to the Democrats, and there were 19 new members on the House Appropriations Committee.[10] These changes meant that a new set of actors were called upon to support SIRTF and the Great Observatories.

Despite the changeover from the Bush to the Clinton administration, Dan Goldin had been asked to stay on as NASA Administrator. Goldin took this as an endorsement of his policies, and "faster, better, cheaper" became the Agency's mantra. Goldin's style and policies did not mesh well with the Old Guard. Charlie Pellerin left in the summer of 1992, a few months after Goldin arrived. Len Fisk left a year later. By November 1993, NASA Headquarters looked quite different to Marcia Rieke and the rest of the SIRTF team. Fisk had been replaced by Wesley T. Huntress, a scientist from Caltech and JPL. Pellerin's position remained unfilled for nearly a year, until Daniel Weedman, a professor at Penn State and a collaborator of Jim Houck's, became the new Director of Astrophysics. Larry Caroff was one of the few infrared-science supporters who stayed at Headquarters after Goldin came to NASA. Weedman and Caroff shared the difficult task of making infrared science a priority in the NASA budget.

SIRTF was out of step with the direction in which Goldin was taking NASA. Pellerin had

---

6. Kwok interview (Session II), 25 March 2009.

7. Pellerin interview, 19 March 2009.

8. Rich Miller, "Meiosis Study for SIRTF," reproduced in the SIRTF SWG Meeting Minutes, 1–2 December 1992, Appendix O.

9. Pellerin interview, 19 March 2009.

10. Memorandum to Space Science Working Group on House Reorganization, dated 18 November 1992, included in the SIRTF SWG Meeting Minutes, 1–2 December 1992.

been right: Shrinking SIRTF by 30 percent to fit on an Atlas rocket was not enough. Chipping away at the edges would not make SIRTF sellable. Headquarters had radically changed. SIRTF would have to do the same.

At times it seemed that the SWG was the only group still thinking about SIRTF since the project had moved to JPL. Pellerin was no longer at Headquarters, which reduced the drive to complete the Great Observatories. Headquarters had its hands full with fixing Hubble, while JPL was focused on problems with the Galileo and Mars Observer missions. While JPL was putting out fires, rumors erupted that Goddard might be angling for SIRTF.[11] It was certainly a plausible story; now that their work with Hubble was drawing to a close, Goddard engineers were likely scouting for new projects. Goddard was already helping Giovanni Fazio develop IRAC, one of SIRTF's three scientific instruments, so why not the build the spacecraft and manage the mission, too?

The SOFIA and Edison Threats

Whether or not Center management of SIRTF was actually in dispute, fights over funding were very real. Groups were competing for a slice of a shrinking pie. Should money ever become available for a New Start, two projects were lining up to eat SIRTF's lunch: SOFIA and Edison.

SOFIA, the Stratospheric Observatory for Infrared Astronomy, was an airborne infrared telescope being developed at Ames, and Edison was a space-based infrared telescope being developed by Scottish and American researchers. Both projects were being pitched as cost-effective, next-generation infrared instruments. Although the scientific community supported SIRTF, it was anxious to get a new infrared instrument up and running to follow up on the findings of the COBE and IRAS missions of the 1980s. No one knew if or when SIRTF would fly. SOFIA and Edison provided a way to hedge one's bets.

The advocates for SOFIA were in many cases the same people who advocated for SIRTF, who naturally used many of the same arguments. In fact, Marcia Rieke was asked by Larry Caroff to set up a joint SIRTF/SOFIA advocacy program.[12] While sharing such activities was efficient, it also blurred the distinct differences between the two projects, making it unclear to OSSA director Len Fisk why both projects were needed. This resulted in the SIRTF and SOFIA teams arguing over which project was better, more important, and further along in development. Despite some overlap, they differed in one critical way: SIRTF had greater sensitivity for seeing fainter objects, whereas SOFIA had higher resolution for seeing brighter objects. Like a wide-angle versus a close-up camera lens, SIRTF and SOFIA were tools equally valuable to astronomers for their research.

Weedman and Caroff had to referee these arguments, and whichever project they supported, the other project would argue against. "These are the kind of tightropes, the kind of juggling, that has to be done at Headquarters, which makes the Headquarters job so challenging on the one hand and so invigorating on the other," Weedman later recalled. "You're juggling all of these competitions. The Centers are competing with each other. The different segments of the community are competing with each other. So you have to find a path that accommodates all these competitions and gets it done in the end. That's why the management challenge is so complicated."[13] There weren't enough funds

---

11. "IR/SubMM/Radio Branch, Book IV, March 15, 1990, to September 11, 1990," Lawrence J. Caroff notebooks, PP08.18, Box 1, Folder 5, NASA Ames History Office, NASA Ames Research Center, Moffett Field, CA.

12. Manning and Caroff interview, 3 September 2008.

13. Daniel W. Weedman, interview by author, Washington, DC, 27 May 2009.

to support both SIRTF and SOFIA in their current configurations. Complicating matters, the Bahcall report was so ambiguously worded that both project teams believed their own project had received highest priority—SIRTF as the highest priority New Start for large projects and SOFIA for moderate-sized projects.[14] Unfortunately, each project was estimated to cost a billion dollars, and NASA couldn't support both. The Bahcall report failed to provide NASA with a clear priority or political cover for canceling one over the other. To resolve the conflict, the science teams pushed to have the Bahcall committee review the two programs again and render a definitive opinion. Both project teams probably thought they would win. However, Caroff thought both would lose, as weakening support for one project would inevitably weaken support for the other.[15] The chairman of the National Research Council's Space Studies Board, physicist Louis Lanzerotti, reluctantly agreed to do a limited review.[16] The result was an even more carefully worded report asserting that both SIRTF and SOFIA were needed.

With the backing of Lanzerotti's report, Weedman and Caroff sought to eliminate the either/or perception regarding SIRTF and SOFIA.[17] Weedman worked on the policy makers, while Caroff focused on the project teams. Caroff pushed to lower the cost of both programs by reducing their capabilities, particularly where they overlapped. Thus the differences between the programs would be clearer, and the savings might make it possible to afford both as smaller programs. By the time Caroff left Headquarters to return to Ames in 1997, both projects had been scaled back in ways that made them complementary rather than competitive missions, and able to fit together within the NASA budget. "Dan [Weedman] was instrumental in getting SIRTF sold," Caroff recalled. "He took over the division at the critical time for a year or two, and he got SIRTF and SOFIA sold—[a] master strategist."[18]

The other project that was making life difficult for SIRTF was Edison, a large infrared space telescope concept initially conceived by Tim Hawarden, of the Royal Observatory at Edinburgh, Scotland.[19] It was potentially a "faster, better, cheaper" mission because it needed no cryogen, instead using the cold temperatures in space to passively cool the infrared instruments. This was a radical departure from all other infrared space missions—IRAS, COBE, Fazio's IRT on the Shuttle, and the upcoming European Space Agency's ISO project—which had immersed the entire telescope, mirrors and all, in a cryogen bath called a dewar. Analogous to a thermos, the dewar was a proven technology, which SIRTF was also planning to use. In contrast, Edison eliminated the cryogen entirely. This dewar-free approach had never been tried, but if it worked, it would reduce the weight, complexity, and cost of the mission.

Advocates for Edison were already involved with NASA through the SOFIA project, including astrophysicist Harley Thronson of the University of Wyoming and Daniel Lester of the University of Texas at Austin. After funding Hawarden's initial work the European Space

---

14. Bahcall report, pp. 75–80.
15. Manning and Caroff interview, 3 September 2008.
16. Letter from Louis J. Lanzerotti, chair of Space Studies Board, to Wesley T. Huntress, Associate Administrator for Space Science, NASA, dated 21 April 1994.
17. Weedman interview, 27 May 2009; also Caroff interview, 3 September 2008.
18. Manning and Caroff interview, 3 September 2008.
19. Hawarden's original project was called "Passively-cooled Orbiting Infrared Observatory Telescope" (POIROT); see H. A. Thronson et al., "The Edison Infrared Space Observatory," *Space Science Reviews* 74, nos. 1–2 (1995): 139–144.

Agency decided not to pursue it further, possibly because ISO was consuming its resources and attention. Thronson, who had worked with Hawarden while on a sabbatical at Edinburgh, wanted to pursue the Edison concept when he returned to the United States and thought that NASA might be willing to develop it.[20] The idea of passive cooling was promising, but it needed much more engineering analysis to see whether it was feasible. Such an analysis seemed harmless and prudent, so Caroff provided Thronson with a little money out of his discretionary funds at Headquarters to get some studies under way.[21]

On the shoestring budget NASA provided, Thronson was able to hire two people (working during their free time) to calculate the thermal profile and optical sensitivity for a parabolic, passively cooled telescope design: Ramona Cummings, at Marshall, who worked on the engine bells for the Space Shuttle; and one of Thronson's engineering graduate students ("a computer whiz") at Wyoming, whose master's thesis was designing an improved snowplow. They weren't experts in space telescopes, but they knew how to model parabolas. Interestingly, both engineers were surprised to find that the temperature inside a parabolic curve would drop to somewhere between 10 K and −40 K. Thronson was not surprised: Tim Hawarden, "with his HP calculator, came up with this number a couple of years ago," he remembers.[22]

These early results were promising, so the Edison team ran with it. Thronson and his colleagues tried to generate awareness of the concept through journal publications and presentations at academic conferences. "Present your stuff in public, let folks throw rocks at it," Thronson says. "You hope that enough folks will read it and they'll say, 'Yes, you know, I think these guys are right.' The SIRTF people did basically the same thing in the early '80s, about a decade earlier."[23] The advocacy for Edison did not go over well with people working on SIRTF or at Headquarters. Caroff recalled:

> [Edison] rose up and people gave it a lot of attention, and they were advocating it to Congress and at the Agency and saying that it could do the job of SIRTF plus a lot more, a lot cheaper, and would last longer.... There was an enormous row. I mean, the SIRTF people were completely—Oh, they were so up in arms and angry it wasn't even funny.... Here are the people at NASA asking Congress to spend a billion dollars on this infrared mission [SIRTF]. Here is [Edison], a bunch of other people coming along saying, "Pfft, this whole thing can be done for a couple hundred million. We can put this thing together and carve it out of balsa wood and launch it with a rubber band." They have not studied the thing in anywhere near the depth that SIRTF had been studied. They hadn't done the engineering studies. They were just doing sort of back-of-the-envelope, order-of-magnitude calculations... But the idea that they would trump this thing, when SIRTF was in a delicate part of its life trying to get sold—basically what they are saying, or you could interpret what they were saying, was 'SIRTF's old hat, we've gone beyond that, why would you want to put money into that when you can do this?' Here is a bunch of people on SIRTF who have invested [almost 20] years, and a center that's invested an

---

20. Harley A. Thronson, interview by author, Long Beach, CA, 4 January 2009.
21. Manning and Caroff interview, 3 September 2008.
22. Thronson interview, 4 January 2009.
23. Thronson interview, 4 January 2009.

awful lot of its future in this mission. And a lot of people who think this Edison mission is just viewgraphs right now. There is no engineering to back it up. They did [calculations], but there really wasn't an in-depth engineering study of it to back it up. So they were just coming in with these concepts and ideas and challenging SIRTF. Do this debate in the scientific community behind closed doors and stuff like that, fine. Do it out in the open with Congress involved, you know, and people getting wind of the fact that NASA's 'probably doing something really stupid. Why, here are these boys from the Royal Observatory of Edinburgh and a few others, University of Wyoming, they can really teach them a thing or two about how to do missions.' That was the problem."[24]

Administrators at Headquarters and the SWG scientists were worried that Edison could derail SIRTF, and some (but not all) of JPL's engineers wanted nothing to do with an untested approach.[25] The resistance to Edison was overwhelming. "Frank Low probably chewed somebody's tail," Caroff said when describing the pressure exercised by SIRTF's advocates. Caroff himself co-opted the Edison team, redirecting their energies into SOFIA. "I really co-opted Harley [Thronson], because when I left Headquarters, I asked him if he'd take my place. And he did."[26]

"I had been working on various advisory committees and so on for NASA," Thronson said. "Larry [Caroff], like Mike Werner and the other folks, had really sweated to get SIRTF and SOFIA [funded].... Larry had been in that role for ten years, and I think he was really ready to go. The infrared astronomy community owes Larry a lot.... [He] invited me out to dinner, and we had a long talk about where NASA astrophysics was going. When I woke up the following morning, I discovered that I had tentatively accepted a job at Headquarters." Thronson arrived in early 1996 and was put in charge of SIRTF. "The job of the program scientist at Headquarters is to represent the project—its scientific goals—to Headquarters and represent Headquarters' goals for the project to the mission," Thronson said. "I saw to it—as did other folks—that SIRTF had as easy a time as one could have."[27]

SIRTF was not only alive but thriving in 1996 when Thronson took over. To reach that point, it had been radically redesigned. By 1993, when Edison and SOFIA were biting at the heels of the SIRTF team, it took a step back and came up with a mission concept that was unassailable.

---

24. Manning and Caroff interview, 3 September 2008. The passively cooled approach of Edison is being adopted by the James Webb Space Telescope (JWST). This claim is supported by John Mather's Nobel bio, which mentions the link between JWST and Edison (*http://www.nobelprize.org/nobel_prizes/physics/laureates/2006/mather-autobio.html*); the presence of Matt Mountain on JWST's SWG and as a coauthor of a 1990 proposal on Edison (*https://jwst.nasa.gov/meet-mountain.html*); and an article by Edison proposal coauthor John K. Davies in *The Space Review*, 2006 (*http://www.thespacereview.com/article/688/1*), and in the same journal, 1992 (*http://www.springerlink.com/content/w63246064028331x*). Another project, SAFIR (Single Aperture Far-Infrared Observatory), is borrowing some of Edison's far-infrared technology concepts, according to the joint Manning and Caroff interview, 3 September 2008. For more details, see Dan Lester et al., "Large Infrared Telescopes in the Exploration Era: SAFIR" in *UV/Optical/IR Space Telescopes: Innovative Technologies and Concepts III*, SPIE Proceedings 6687, ed. Harold A. MacEwan and James B. Breckinridge (Bellingham, WA: SPIE, 2007). Note that SAFIR, which Caroff pronounced as "sapphire," is not the same as SAFIRE (Submillimeter And Far Infrared Experiment), an instrument on SOFIA.

25. Dr. Donald Rapp, who for many years was chief engineer of missions, tried to find a way for JPL to be involved (personal communication with author, 19 November 2007 and reports).

26. Manning and Caroff interview, 3 September 2008.

27. Thronson interview, 4 January 2009.

A Retreat

In the hope that the SWG might develop some new ideas for SIRTF, Jim Houck suggested that they schedule a planning retreat. They chose a weekend in November 1993 to convene in Broomfield, Colorado, near the headquarters of Ball Aerospace. Ball had been a longtime partner on SIRTF on various engineering studies and most recently had built some of the corrective instrumentation for Hubble. The meeting in Broomfield brought together the SWG and the instrument team members. The primary goal of the meeting was to come up with a SIRTF that would sell.[28] The feeling was that if they didn't come up with something new, that could be it for SIRTF. The program might be permanently canceled.

The project design stood at $1.3 billion, which was an improvement over earlier designs. While SIRTF was still heavy enough to require an Atlas rocket, it was no longer so heavy that an expensive Titan would be needed. The telescope's three instruments—the near-infrared camera, the far-infrared camera, and the spectroscope—had a range of capabilities that appealed to a broad set of scientists. The Atlas-SIRTF concept had been well received in a presentation given earlier that summer by then-project manager Jim Evans to his boss, George Newton, who oversaw SIRTF and other programs at Headquarters.

Since then, the rules and project manager had changed. Larry Simmons, officially SIRTF's new project manager, was back at JPL wrapping up his responsibilities for the Hubble repair mission. After headline-making losses, NASA wanted more wins and smaller bets. "Goldin decreed that he wanted to fill the sky, blacken the sky with satellites," Simmons recalls:

You can't do that if every one of them costs several billion dollars. So he said, "We're not going to launch any missions that cost more than a half a billion dollars." So Jim Evans' billion-dollar SIRTF was sort of dead on arrival; so Jim went on to another project. And I had just finished building the camera that fixed the Hubble.... I had a certain reputation at NASA for getting tough things done. I think that contributed, frankly, to why I was asked to take over SIRTF. The guideline I was given was that we could do anything we wanted, but the project had to cost less than a half a billion dollars. That was a starting point. And that had to include everything—launch services and all of that.[29]

For 20 years, SIRTF's advocates had done their best to accommodate the demands of Agency administrators, whether by accepting new Center management, flying on a Shuttle, or conducting "bring me a rock" studies. This request to cut the budget by half after the project had already been de-scoped left many of the scientists wondering if SIRTF would still be worth doing at such a price. To the three instrument PIs, it just might be OK, as long as it wasn't *their* instrument that got cut.

Balancing tradeoffs is part of the management job. Simmons transitioned to SIRTF after leading the Hubble repair mission. As a new member of the team, says Simmons, "one of the things I had to do was find out what it was [the scientists] wanted":

What they wanted was to get really interesting infrared pictures of the sky, ultimately. So how were they going to do that? Well, it turned out they didn't all have the same vision. Some of them were interested in the

---

28. The description of the Broomfield meeting in this chapter is based on Rieke interview, 9 June 2009; Werner interview, 15 December 2008; Simmons interview (Session I), 18 February 2009; and Gehrz interview, 5 January 2009.
29. Simmons interview (Session I), 18 February 2009.

far infrared, some of them were interested in the near infrared. Some were interested in spectroscopy, and others were interested in imaging. So there were all kinds of interests. It was interesting being the new guy. Some of them would take me aside, one-on-one, and say, "You don't have to worry about that guy over there, he doesn't have very much influence on things anyway." So I'd listen to him, and then the other guy would come along and say, "What he's trying to do isn't really going to work, and it's just going to sink this ship, so don't pay too much attention to him." So I realized soon that I couldn't just go to one guy and say, "What should we do?" I really had to understand the team.[30]

Simmons, like Mike Werner, had a very open management style. He was willing to share financial information—how much was going to whom, and for what purpose—with all members of the team, whether they were from NASA, the academic community, or private industry. This openness helped the team to understand the financial impact of their actions on the project as a whole.

At Broomfield, Mike Werner tried to get the SWG to focus on the bigger picture, too. Rather than getting caught up in squabbles about who was going to lose what functionality on their instrument, he sought to build consensus around the science that SIRTF needed to accomplish. The PIs had invested years in engineering their instruments, but they were scientists above all. Werner counted on them to recognize that their instruments were being built in service to the scientific objectives.

And what were those objectives? The Bahcall report, which indicated that SIRTF was the most important (large) mission, also laid out the core scientific questions in infrared astronomy. It was hard to argue with a document that presumably represented the consensus of the infrared community, of which the SWG were all members. Werner and his deputy project scientist and right-hand man, Tom Soifer, tried to identify science topics that matched those of the Bahcall report. Mike Jura, George Rieke, and Bob Gehrz also helped to refine the list. Thus, with relatively little argument, the PIs—Rieke, Giovanni Fazio, and Jim Houck—and the rest of the SWG achieved consensus on the questions that the suite of instruments needed to address. The four science goals for SIRTF were to observe 1) protoplanetary and planetary debris disks (essentially dust clouds that are in the process of coalescing into planets), 2) brown dwarfs and super planets (intermediate structures between solid-core planets and fusioning-core stars), 3) ultraluminous galaxies and active galactic nuclei (in which star formation and black holes might be observed), and 4) the early universe (deep surveys to capture emissions from the distant edges of the universe).[31]

These four scientific programs formed the yardstick against which tradeoffs could be measured. A feature could not be cut out if it impaired (or added if it failed to enhance) SIRTF's ability to gather data needed to address the four areas. This simplified the mission, in both cost and complexity.

After deciding on what sort of data SIRTF absolutely must be able to gather, the SWG turned to the question of how to bring down costs. The main ways to accomplish this are to either reduce the launch weight or increase the

---

30. Simmons interview (Session I), 18 February 2009.

31. Michael Werner, "A Short and Personal History of the Spitzer Space Telescope," 1995, ASP Conference Series, preprint available at *http://arxiv.org/PS_cache/astro-ph/pdf/0503/0503624v1.pdf* (accessed 30 August 2016); and SIRTF SWG Meeting Minutes, 16–17 December 1993.

viewing efficiency. SIRTF was heavy because it needed to carry sufficient cryogen (a few thousand liters) to ensure the instruments would be sensitive enough for infrared observations and to give them a lifespan long enough to make those observations. SIRTF's efficiency depended on how long it had a clear view of space, without radiation from Earth, the Moon, or the Sun.

As noted, for the Atlas-launched mission SIRTF was now 30 percent lighter than it had been for the Shuttle-launch mode, allowing a launch into high-Earth orbit. It had been redesigned for operation 100,000 kilometers above Earth, for which the orbital mechanics were well understood. Johnny Kwok's innovative proposal—to place SIRTF in an Earth-trailing solar orbit—was untried, but it had great promise as a way to increase the telescope's viewing efficiency. The SWG decided that the risks were worth it if they could maximize observation time.

With clear science goals, and an orbit that optimized efficiency, the remaining challenge was to bring down the weight. One night, Frank Low hit upon the idea of doing a warm launch. Typically, infrared telescopes are put into a dewar of cryogen and cooled to a few kelvins while still on the ground. When the telescope reaches orbit, it's ready to go. Low suggested putting only the infrared instruments—not the mirrors and other optical components—into the dewar and letting the other components cool in space. It might take a few weeks for the telescope to radiatively cool in orbit, but this change meant a substantial reduction in the quantity of cryogen that would need to be carried aboard the Atlas and a corresponding weight savings at launch. Alternatively, the same amount of cryogen could be made to last longer, increasing SIRTF's operational lifespan.

The warm-launch concept was a breakthrough. It reduced both the weight and the cost of SIRTF without compromising its capabilities. The Ball engineers, who had built dewars for all of the previous infrared missions, agreed that putting only the instrument chamber, rather than the whole telescope, into a cryogen dewar just might work. They were willing to give it a try.

As with any great idea, there are arguments about who deserves credit for the warm-launch concept. Frank Low indisputably brought forward the idea that the SWG adopted in 1993. However, the notion of passively cooling a telescope in space had been advanced by Tim Hawarden in 1989. The Edison team had developed the concept further but may have gone too far by suggesting that no cryogen at all was needed for an infrared telescope. Even with radiative cooling, SIRTF would have to be cooled to operate at the far-infrared wavelengths. And another group, led by Goddard's Harvey Moseley, recommended a similar passive-cooling approach in their proposal during the SIRTF center competition in 1989.[32] The historical record cannot resolve the debate, and it seems quite likely that Hawarden, Moseley, and Low independently developed their ideas.[33] However, they did so in an environment that they shared, and it seems that any credit should probably be shared as well.

With a warm launch, SIRTF would be light enough to launch on a Delta rocket. The Delta

---

32. In interviews, Larry Manning, Larry Caroff, and Giovanni Fazio reported that S. Harvey Moseley had a similar idea (Manning and Caroff interview, 3 September 2008; Fazio interview, 26 May 2009).

33. Sadly, both Hawarden and Low passed away in 2009. However, their ideas for radiative cooling (and those of Moseley, who oversaw the construction of SIRTF's IRAC instrument) live on in the design for the James Webb Space Telescope, the follow-on mission to SIRTF and Hubble that is being planned for launch after 2017. Their accomplishments were recognized by their peers: Hawarden posthumously received the NASA Exceptional Technology Achievement Medal (2010). Low received the NASA Exceptional Public Service Medal (2008) and the American Astronomical Society's Joseph Weber Award for Astronomical Instrumentation (2003). Moseley received the Weber award (2007).

TABLE 7.1. Chronological changes to SIRTF (compiled by Johnny Kwok).

Period	1972–1984	1984–1988	1988–1992	1992–1994	1994–1996	1997
Orbit	300 km	900 km	70,000–100,000 km	Solar	Solar	Solar
Launch Vehicle	STS	STS, OMV	Titan IV	Atlas II-AS	Delta II-7920	Delta II-7920H
Tracking System	TDRSS	TDRSS	DSN 26M	DSN 34M	DSN 34M	DSN 34M
Mass (kg)	4500	5560	5500	2460	750	865
Lifetime	30 days/launch (serviceable)	2 years/launch (serviceable)	5 years	3 years	2.5 years	5 years
Primary Mirror (cm)	100	95	95	85	85	85
Helium (liter)	350	6600	4000	920	250	350
Wavelength	2–2000	1.8–700	1.8–700	2.5–200	3.6–160	3.6–160
Detectors	n/a	~10,000	~50,000	~140,000	~350,000	~350,000
Cost	Godzillion $	Godzillion $	~$2B	$860M	<$500M	$450M*

The mass increase from 1996 to 1997 was due to the launch vehicle being upgraded to a Delta 7920-H, which provided additional launch capability that was used to increase the cryogen and, in turn, SIRTF's operational lifetime (SIRTF SWG Meeting Minutes, 16–17 December 1996, archived with NASA Headquarters History Program Office). Conducting cost estimates prior to 1988 was difficult because SIRTF's performance specifications had not yet been established. As a result, estimates were unreliable until the late 1980s, when the SWG (formed in 1984) developed a set of specifications for SIRTF that met scientific needs. Budget exercises prior to that time, such as those recorded in the Phase A Statement of Work, reveal multiple approaches for estimating costs, including taking IRAS as a template and applying a 3x factor for increased complexity, resulting in a cost estimate for SIRTF of ~$130 million (1980, real dollars). Note that this did not include launch costs or integration costs with the Shuttle, and in regard to costs, IRAS could provide little guidance, as it was launched by rocket (Statement of Work Specification, Phase A, Space Infrared Telescope Facility, AFS1070.8A, Archives Reference Collection, FC5:D4, NASA Ames History Office, NASA Ames Research Center, Moffett Field, CA).

* The total cost of SIRTF was $708 million (in 2001 dollars), plus $68 million for launch services. If SIRTF had launched in Dec. 2001 as planned, the telescope cost would have been $515 million. Launch delays contributed to the increased cost, but also enabled further development and readiness testing that has led SIRTF to far exceed its planned usefulness. As of this publication, data collection is expected to last until mid-2018. Details are from personal communication with Mike Werner, Feb. 24, 2017.

was smaller and therefore cheaper to launch than an Atlas rocket; but because the Delta's payload area was limited, SIRTF's mirror could not exceed 3 feet in diameter. If SIRTF could be made to fit, the cost savings would make the mission affordable. SIRTF had gone from a $1.2 billion, 5,700-kilogram Titan-launched concept to a $400 million, 750-kilogram Delta-launched concept.[34] It was now looking more like the IRAS telescope, which had also launched on a Delta rocket. Although SIRTF's detectors were more sensitive, its orbit was more efficient, and its design made better use of the cryogen, the cosmetic resemblance between SIRTF and IRAS was strong (Table 7.1 and Fig. 7.1). Although the trailing-Earth orbit and warm-launch schemes were untested, the cost profile and similarity to a successful mission balanced the perceived risks and, as the director of JPL put it, made SIRTF "exactly the kind of innovative mission" that NASA wanted.[35]

---

34. SIRTF SWG Meeting Minutes, 9–10 May 1995.

35. SIRTF SWG Meeting Minutes, 9–10 May 1995.

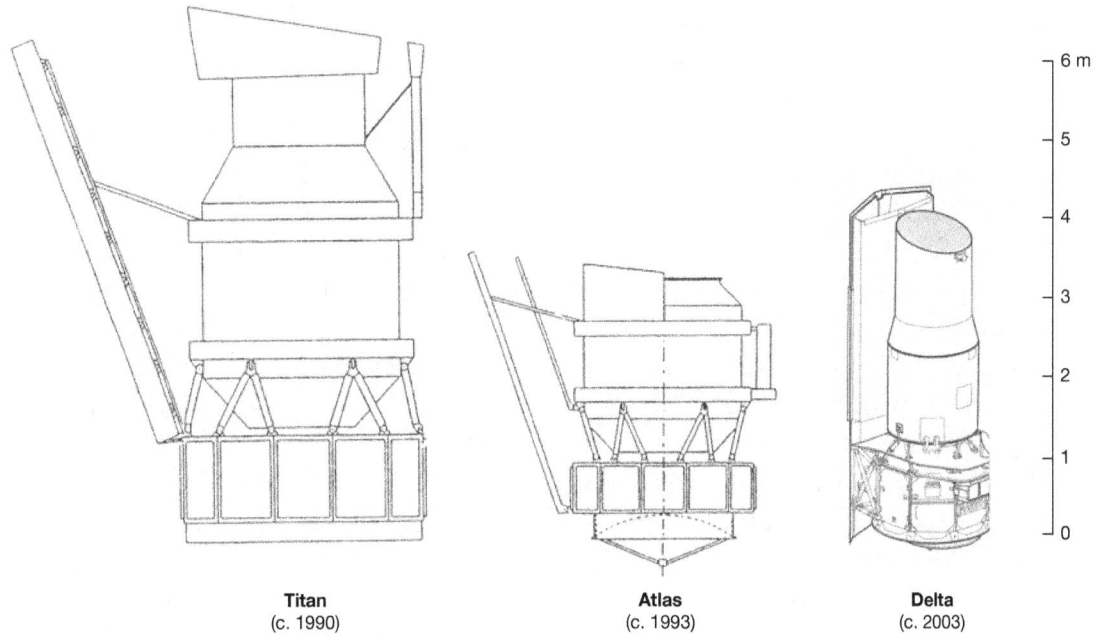

FIGURE 7.1. The evolution of SIRTF as a free-flyer: Comparison of models launched by Titan, Atlas, and Delta rockets.

Poster Child

The Broomfield meeting was a huge success, as it pulled together the three key elements—clear scientific goals, an innovative orbit, and a warm launch—that made SIRTF not just sellable but also a poster child for "faster, better, cheaper." Yes, faster. A month after the Broomfield meeting, Mike Werner presented the new SIRTF concept to Dr. Richard Obermann, science advisor to the House Subcommittee on Space and Aeronautics.[36] By April 1994, the concept had been sufficiently developed to win over JPL management.[37]

SIRTF, for the first time in its history, appeared in the national budget. Congress appropriated $10 million in the FY96 budget specifically for SIRTF. This was short of the $15 million Headquarters had requested, but, as Weedman said, "The amount of money was irrelevant, because what NASA wanted was Congress's blessing of SIRTF, which any allocation achieved."[38] Headquarters added another $5 million from discretionary funds to make up the shortfall. SIRTF was now in Phase B, where the feasibility analyses of Phase A (1984–1995) were converted into an integrated systems design that included detailed engineering specifications, budgets for all aspects of the project, and an operations plan for when the telescope was in orbit. The transition from Phase A to Phase B is perhaps analogous to the condensation of gas into a liquid—with pressure, things become more tangible, more visible, but still fluid. The transition from Phase B to Phase C

---

36. Michael Werner, SIRTF Reconfiguration, presentation to Dr. Richard M. Obermann, science advisor, House Subcommittee on Space and Aeronautics, 10 December 1993.
37. SIRTF SWG Meeting Minutes, 5–6 May 1994.
38. SIRTF SWG Meeting Minutes, 12–13 September 1995.

is analogous to liquid becoming a solid: As pressure increases, things become aligned and freeze into a stable configuration. Phase D involves the preparations for launch, while Phase E includes all mission operations in space.

On 19 November 1996, with funding and paperwork in place, JPL announced that it had signed a Programmatic Commitment Agreement with Headquarters, formally authorizing the design of SIRTF to begin.[39] Lockheed and Ball were brought on as subcontractors to help with the design in Phase B (with additional support from Hughes), and it was expected that they would be retained to build SIRTF whenever construction began in Phase C or Phase D. No one was sure whether that day would come, and the task at hand was to design SIRTF to fulfill its scientific promise and show that it could be done for under $500 million. That was no small challenge, as $500 million was nearly the amount spent on the mission to correct Hubble's mirror. Simmons, who had led that repair effort, was now the SIRTF project manager. He had to manage with a lean budget and without sacrificing the very things that made SIRTF sellable—clear scientific goals, an innovative orbit, and a warm launch, the latter two of which were untried. As Simmons remembers it:

> There were those who said that idea [the warm launch] was crazy and couldn't be done, and there were others who said it *could* be done. I was smart enough to not take sides but to listen to the two sides argue it out. As a matter of fact, there were people on the project—key people on the project—who as late as launch [August 2003] still said we could have done it the other way [a cold launch, with the whole telescope in a dewar]. There were those who just never let go of the fact that we could have done it the other way. I don't believe that. The reason I don't believe that is that we would never have been given the funding to start if we had tried to do it the other way.... Politically, it could not [have been done the other way]. The thing that really got SIRTF going was the fact that we showed that we could do things in a different way. We could satisfy NASA's goals of having industry have a significant role. We could convince people that this approach would work, that it was credible. So they kind of dipped their toe in the water and let us go forward with this approach. Another very important part of SIRTF, I should say, was the contribution by the celestial mechanics people to not go into Earth orbit. All the missions previously of this nature had gone into Earth orbit—including Hubble, including IRAS. The mission designer, a guy named Johnny Kwok, is the guy who said, "We could use a smaller rocket to put a bigger payload into an orbit going around the Sun instead of going around the Earth." It takes about a microsecond to figure out that there are some real advantages to that. One of them is that you don't have to go through the heat-and-cold cycle once a day. The thermal environment is absolutely constant, because you're a constant distance from the Sun. The trick was to come up with an orbit—a solar orbit, it's called—that didn't separate [SIRTF] from the Earth too fast. If it went too fast away from the Earth, then communications would be a problem, because the $1/R^2$ problem—the farther away you get, the more power you need to transmit signal. We wanted to send fairly high data rates, so we wanted communications. The celestial mechanics guys, the orbit guys, figured out how we could come up with a way to get the

---

39. SIRTF SWG Meeting Minutes, 16–17 December 1996.

maximum amount of material into space, into the solar Earth-trailing orbit, that allowed for a telescope that would be very stable, could make lots of observations, and could operate 24/7—which was something I think other people really envied when they saw what we could do. So these pieces kept coming together, and we kept building on this concept. Pretty soon, people started hearing about it, and saying, "Oh, come tell us about it." It became interesting to both the science community and the management community at NASA Headquarters to hear what we were going to do with this.[40]

---

40. Simmons interview (Session I), 18 February 2009.

CHAPTER 8

Constructing SIRTF

SIRTF's Phase B design stage lasted only two years. Compared with the difficulties in Phase A in becoming (and remaining) one of NASA's priorities—from releasing the Announcement of Opportunity in 1983 to getting congressional approval for funding in 1996—Phase B was relatively easy. The team moved quickly through this phase, in which the SWG weighed scientific and technical tradeoffs, consolidating the knowledge gained from earlier studies and tests. Project scientist Mike Werner described the Phase B period:

> Once we got the authority to go ahead, then things started happening a lot quicker. There'd been an in-house design activity led by a guy named Jim Fanson, ... a very, very talented engineer, [who] really brought the warm-launch concept to life. Then we went out and selected contractors, not to build a mission to a set of specifications but to help us implement ... the warm-launch concept.... [People from] the instrument teams [and] our facility scientists, Frank Low and George Rieke and Tom Roellig, particularly Tom, spent a lot of time at JPL interacting with the contractors, making sure they knew what their jobs were, defining interfaces, roles and responsibilities—who builds this part, who builds that part, and so forth. After that, we quickly got into a regular cadence of quarterly reviews, monthly management meetings, frequent teleconferences, interactions with the contractors.[1]

During this period, Werner made sure that the team stopped once in a while to celebrate milestones and breakthroughs.

SIRTF did not remain long in Phase B because the feasibility of the design was readily established. The detectors, which were critical to the success of the design, had benefited from more than two decades of development by the SIRTF team.[2] The cryogenic system had been largely demonstrated by the performance of IRAS, whose launch in 1983 unleashed a flood of scientific results into the academic journals. Prototypes of other challenging components,

---

1. Werner interview, 15 December 2008.

2. In addition to the SWG members, many people were ultimately involved in the effort to characterize the various semiconductor materials and develop the detectors. A few major contributors should be acknowledged: Craig McCreight (Ames), Harvey Moseley (Goddard), and Judith Pipher (University of Rochester). See Appendix A for a full list of participants.

such as a demonstration mirror successfully milled from beryllium, also helped to establish the design's feasibility. Sometimes prototypes helped in unintended ways. On a visit to Washington, Jim Houck, PI of the infrared spectrograph, brought along a spare instrument housing to show his colleagues at NASA Headquarters. He happened to still have the housing with him when he later met with a congressional representative who was unenthusiastic about funding another telescope after Hubble. The spare was the result of a machining error, but the size and weight were approximately correct—about that of an empty shoebox. "I pulled out the model," Houck remembers, "and he looked at it, and he said, 'That's what it looks like?' And I said, 'Yeah.' He said, 'I'll try to get you a little money.'"[3] The prototypes were useful to show that SIRTF was doable—scientifically and technically, of course, but also economically and politically. At SIRTF's conception in 1971, it had been impossible to build infrared detector arrays and large beryllium mirrors, let alone cool them to a few kelvins in space. By the late 1990s it was no longer impossible, just exceedingly difficult.

SIRTF's Design

The flight-ready design was driven largely by economic and scientific needs. NASA, with Congress's support, had budgeted $400 million to build SIRTF and $300 million to operate it after launch.[4] In return for their public investment, the nation's taxpayers would acquire an instrument capable of addressing major scientific questions about the emergence and evolution of stars and planets.[5] Specifically, SIRTF was being designed for research centered on the four key topics mentioned in chapter 7: a) protoplanetary and planetary debris disks, b) brown dwarfs and super planets, c) ultraluminous galaxies and active galactic nuclei, and d) the early universe.[6] As noted, these four goals were aligned with scientific priorities listed in the decadal survey for the 1990s (the Bahcall report) and fully exploited the unique capabilities of a highly sensitive, space-borne infrared telescope.

To achieve its key scientific goals, SIRTF relied on three instruments that covered the infrared spectrum from 3.6 to 160.0 microns. Two of these—the Infrared Array Camera (IRAC) and the Multiband Imaging Photometer for SIRTF (MIPS)—were cameras that could detect sources emitting in the infrared, while the third, the Infrared Spectrograph (IRS), was a spectroscope, which could detect the sources' chemical composition. These were the same three instruments that had been selected following the 1983 Announcement of Opportunity; however, their realized operating range had narrowed since then. The wavelength coverage was 3.6–160 microns for imaging, 5.3–40 microns for spectroscopy, and 55–95 microns for spectrophotometry.[7] The far-infrared band was eliminated and the

---

3. Houck interview, 25 May 2009.

4. SIRTF SWG Meeting Minutes, 9–10 May 1995, including presentation by project manager Larry Simmons.

5. For more detailed reviews of the design and operation of SIRTF, see Schuyler Van Dyk, Michael Werner, and Nancy Silbermann, *Spitzer Space Telescope Handbook*, Version 2.1, available at *http://irsa.ipac.caltech.edu/data/SPITZER/docs/spitzermission/missionoverview/spitzertelescopehandbook/* (accessed 30 August 2016); R. D. Gehrz et al., "The NASA Spitzer Space Telescope," *Review of Scientific Instruments* 78, no. 011302 (2007): 1–38.

6. M. D. Bicay and M. W. Werner, "SIRTF: Linking the Great Observatories with the Origins Program," in *Origins*, ASP Conference Series, vol. 148, ed. Charles E. Woodward, J. Michael Shull, and Harley A. Thronson, Jr. (San Francisco: Astronomical Society of the Pacific, 1998), pp. 290–297.

7. Spectrophotometry, as the name suggests, uses a blend of spectra and photometric data to obtain spectral energy distributions (SED), which can be used to classify stars. Data are from *Spitzer Space Telescope Handbook*.

spectroscopic capabilities reduced, but the remaining infrared wavelengths held great promise for scientific discovery.

The entire telescope facility was about the size of a car. When the 95-gallon liquid helium tank was empty, SIRTF weighed 1,877 pounds. It stood just over 13 feet (see Figs. 8.1, 8.2, and 8.3).[8] The top third contained the telescope itself. This portion, which was mostly hollow, was where the photons of infrared radiation would be collected and directed onto the 33.5-inch (85-centimeter) beryllium mirror (see Fig. 8.4, p. 121). The mirror sat directly above the Multiple Instrument Chamber (MIC) (see Fig. 8.5, p. 121), which spanned the facility's middle third. Here the three scientific instruments would record the brightness and intensity of the infrared flux collected by the mirror. To bring the instruments to the necessary operating temperature of 5.5 K, they would sit in a dewar of liquid helium. The bottom third of the facility was the spacecraft bus, which contained the electronics and the mechanisms to point the telescope and send data back to Earth.

One side of the facility would be covered by a solar panel, which would simultaneously draw operating power from the Sun and shield the infrared detectors from its light. The three scientific instruments sat atop 350 liters of liquid helium. For a photo and description of the IRAC, IRS, and MIPS instruments, see Figure 8.6.

FIGURE 8.1. SIRTF during final integration and testing at Lockheed Martin, Sunnyvale, California (Russ Underwood, Lockheed Martin Space Systems).

---

8. Spitzer Fact Sheet, JPL Project Office, Doc. #PM 12-12-03, *http://www.spitzer.caltech.edu/file/97-Fact-Sheet* (accessed 30 August 2016).

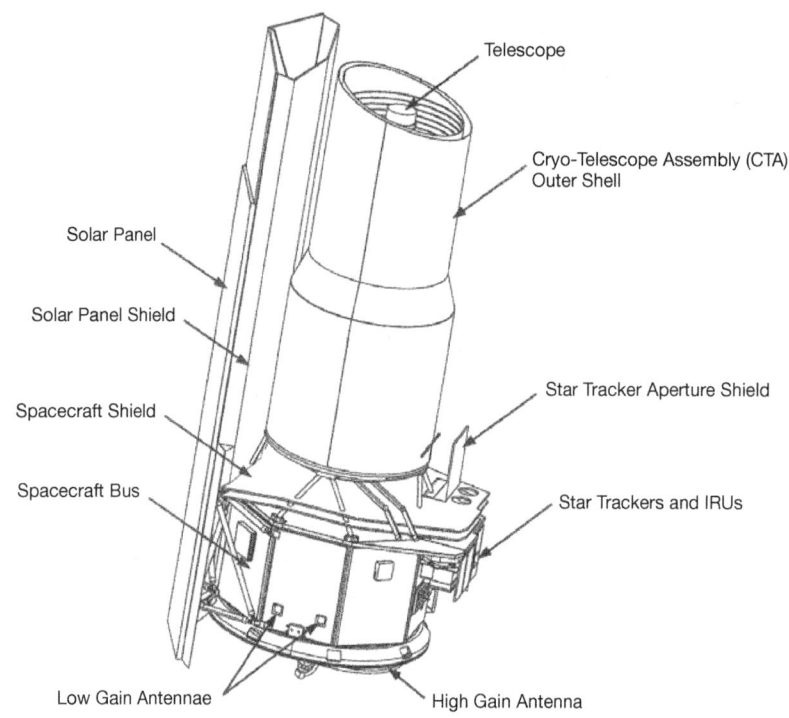

FIGURE 8.2. Exterior components of SIRTF.

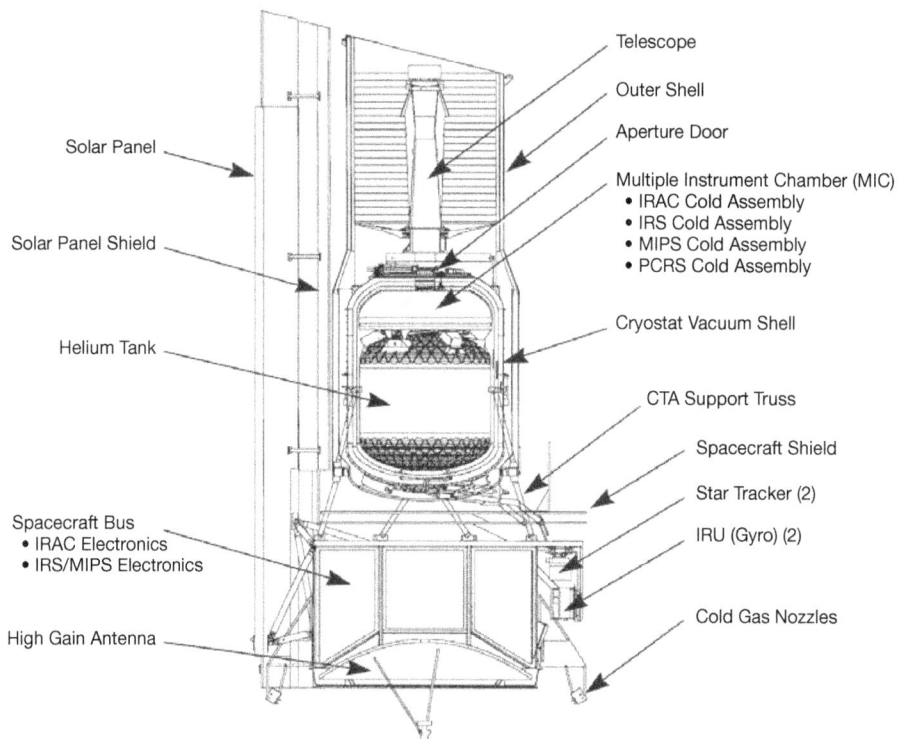

FIGURE 8.3. Interior components of SIRTF.

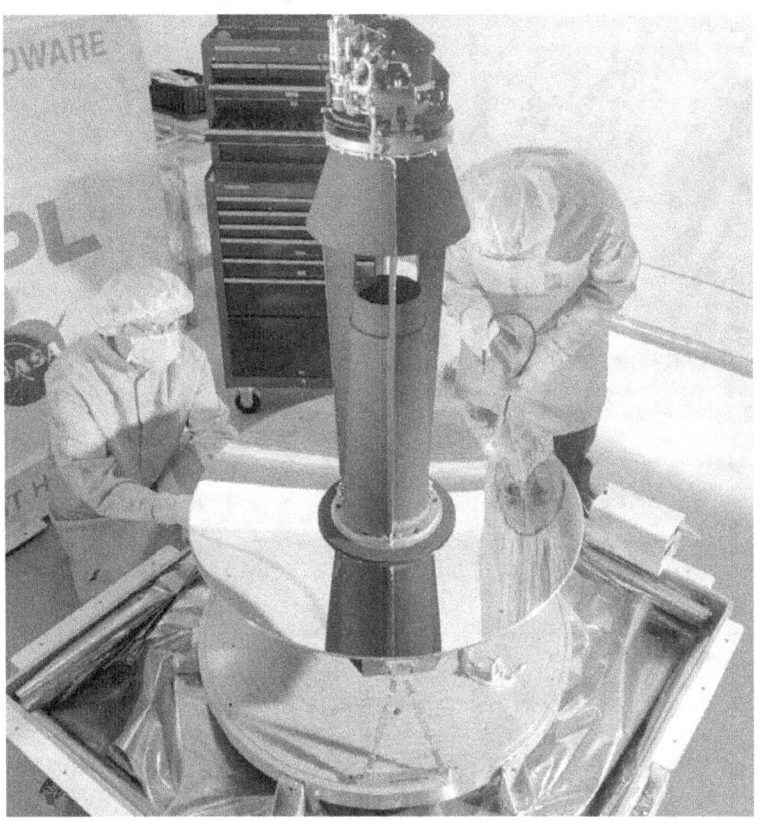

FIGURE 8.4. The beryllium mirror and telescope assembly at Ball Aerospace.

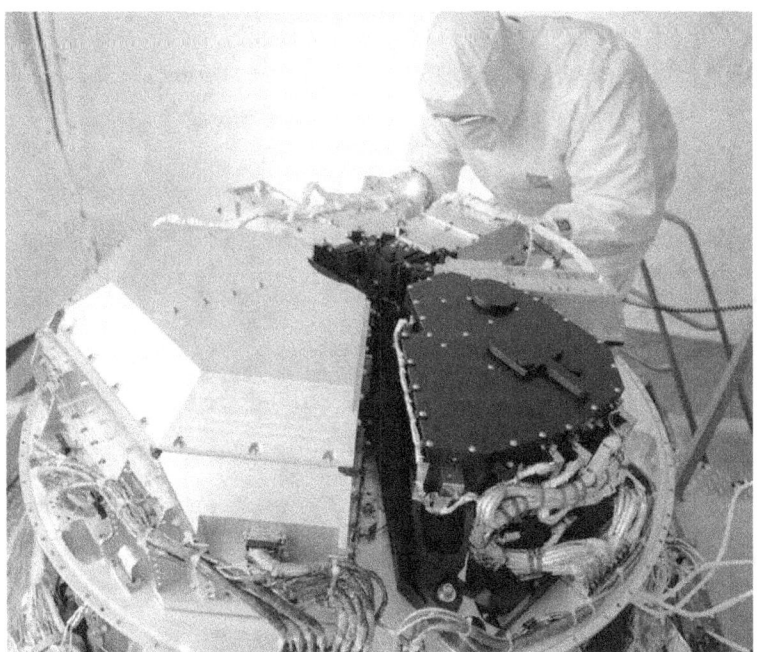

FIGURE 8.5. The three science instruments installed in Spitzer's cryogenic Multiple Instrument Chamber (MIC) at Ball Aerospace.

a) Infrared Array Camera (IRAC)

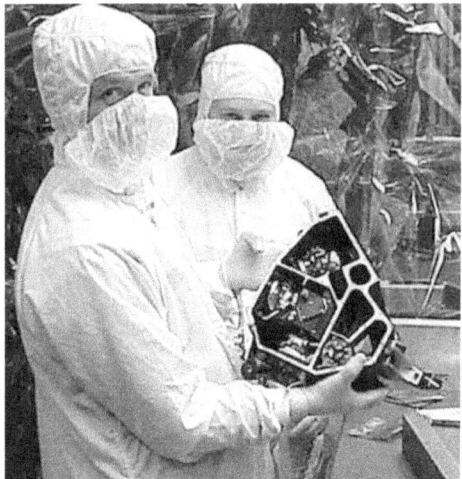

The Infrared Array Camera (IRAC) cryogenic assembly with top cover removed at NASA Goddard, Greenbelt, Maryland.

The IRAC provides large-field imaging in four bands between 3 and 9 microns. The IRAC bands were selected to characterize the starlight from distant galaxies, allowing estimation of their redshifts, and to identify nearby substellar objects (brown dwarfs) by measuring their cool spectral energy distributions.

The near-infrared to mid-infrared imaging by IRAC provides evidence for many of the science themes that have fueled interest in the James Webb Space Telescope (JWST), for which IRAC and SIRTF were important scientific and technical precursors.

IRAC was built at NASA Goddard Space Flight Center (GSFC). Giovanni Fazio (Harvard SAO) is the Principal Investigator. Harvey Moseley (GSFC) is the Instrument Scientist.

b) Infrared Spectrograph (IRS)

The Infrared Spectrograph (IRS) after integration with the Multiple Instrument Chamber at Ball Aerospace, Boulder, Colorado.

The IRS provides low-resolution spectroscopy (4 to 40 microns) that probes the composition of debris systems and the nature of interstellar dust in highly red-shifted galaxies. It also provides moderate-resolution spectra to study the emission lines from infrared-bright galaxies to determine their sources of energy.

The European Space Agency's Infrared Space Observatory (ISO) has shown that the 5- to 37-micron spectral region is rich in both emission lines and broad molecular features. The increased sensitivity of IRS resulting from its high-performance arrays allows these molecular features to be probed to far fainter levels than with ISO.

The IRS was built at Ball Aerospace. James Houck (Cornell) is the Principal Investigator.

c) Multiband Imaging Photometer for SIRTF (MIPS)

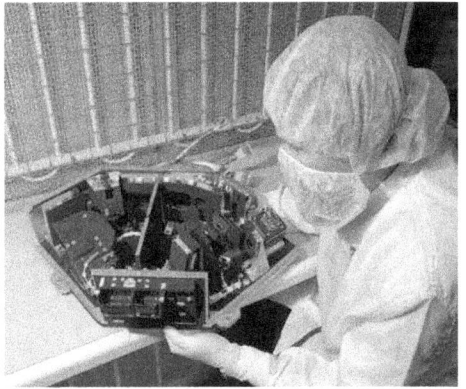

A Ball Aerospace technician holding the Multiband Imaging Photometer for SIRTF (MIPS) prior to integration (Ball Aerospace).

The MIPS supports large field mapping and high-resolution imaging from the mid-infrared to the submillimeter wavebands. Its concept is centered on the study of the far-infrared emissions of distant galaxies due to ultraviolet and visible energy absorbed and reradiated at longer wavelengths by their interstellar dust and on the analysis of the systems of debris around nearby stars associated with possible planetary systems.

The MIPS gives a much deeper look at the far-infrared sky than was possible with either ISO or its predecessor, IRAS, which surveyed the entire sky in 1983. The archive of MIPS data will constitute a fundamental scientific resource for many years.

MIPS was built at Ball Aerospace. George Rieke (University of Arizona) is the Principal Investigator.

FIGURE 8.6. The IRAC, IRS, and MIPS instruments.

A New Start

After two years in Phase B, the SIRTF project office at JPL—working closely with the SWG, instrument PIs, and contractors—had resolved many of the technical issues and was preparing to specify how the system would be integrated and launched and the data retrieved. The final hurdle before building a launch-ready version of SIRTF was to pass the internal Preliminary Design Review (PDR) and the Non-Advocate Review (NAR).

To save time, project manager Simmons was able to schedule the reviews back to back. So, over three days in September 1997, panels of peer reviewers heard from the SWG and the project office on how they planned to build and operate SIRTF. During the first two days presentations focused on the PDR. Those on the final day outlined SIRTF's implementation plan to the NAR board, whose members were not stakeholders in the project. Overall, the project received strong support from these review boards, both of which recommended to NASA management that SIRTF advance to Phase C/D.[9] There was agreement at all levels that SIRTF's scientific goals continued to generate excitement. And costs, reduced by 75 percent from the original 1989 estimate, made the project affordable within NASA's tight budget. As a result, SIRTF was granted New Start status in FY98.[10]

Construction could finally begin. As the sole New Start that year and the de facto flagship mission for NASA, SIRTF was now on a fast track for launch in 2001. A contemporary article by Michael Bicay and Michael Werner explains how the project was organized:

While JPL remains responsible for project management, systems/mission engineering, science management, and flight operations, the other partners have all been identified in the past year and have been actively working together during SIRTF's design phase. Lockheed Martin (Sunnyvale, CA) is the partner responsible for the spacecraft and for the system integration and testing. Ball Aerospace (Boulder, CO) assumes responsibility for the cryogenic telescope assembly and will build the IRS and MIPS instruments. NASA's Goddard Space Flight Center (Greenbelt, MD) will build the IRAC instrument. Finally, the Infrared Processing and Analysis Center (Pasadena, CA) has been designated as the SIRTF Science Center (SSC), and will be responsible for all elements of science operations.[11]

Figure 8.7 (p. 124) illustrates the management configuration. Tom Soifer, who had been Werner's deputy, became the director of the Science Center, and Charles Lawrence replaced Soifer as the deputy project scientist. As of March 1998, 283 people were working on SIRTF.[12] (This figure represents just a small fraction of the total number of people who worked on SIRTF, all of whom are listed in Appendix A.)

SIRTF's Build

The team that had designed SIRTF in Phase B was going to build it in Phase C. This is more unusual than it might appear. Normally, NASA competitively selects several vendors to propose

---

9. SIRTF SWG Meeting Minutes, 29 April–1 May 1997 and 16–17 October 1997.
10. On 25 March 1998, NASA Administrator Daniel S. Goldin authorized the start of work on the Space Infrared Telescope Facility, per JPL Press Release #98-028, available at *http://www.jpl.nasa.gov/releases/98/sirtfgo.html* (accessed 30 August 2016).
11. Bicay and Werner, "SIRTF: Linking the Great Observatories with the Origins Program," p. 295.
12. SIRTF SWG Meeting Minutes, 21–23 April 1998.

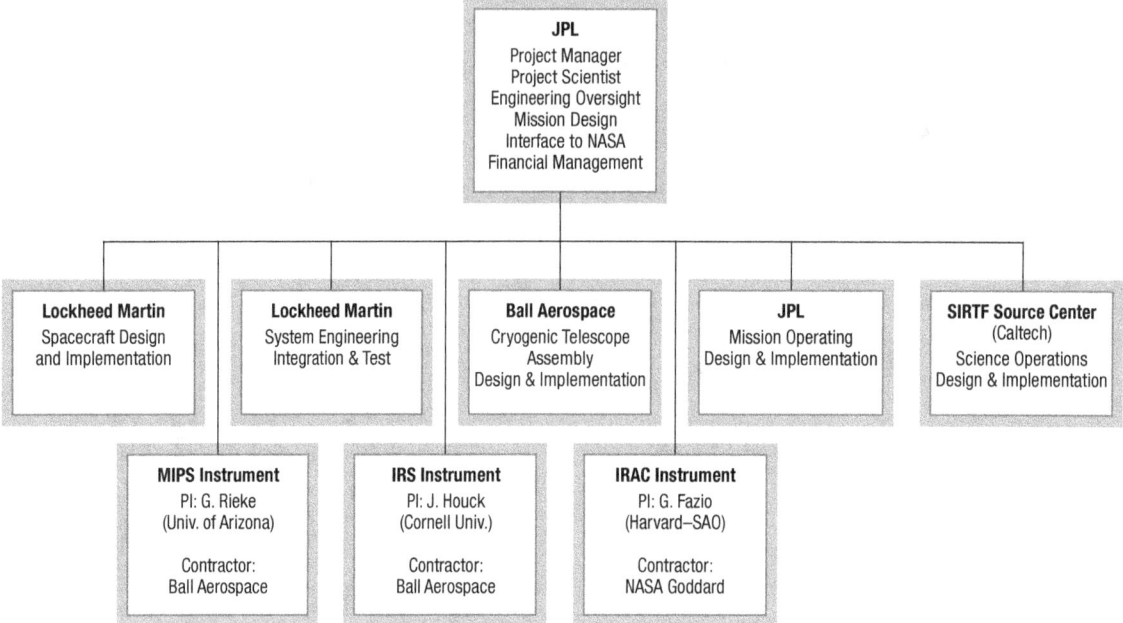

FIGURE 8.7. SIRTF project organization during development (1996–2003).

designs in Phase B, then reopens the competition to select the best vendor to implement the design in Phase C. Instead, the SIRTF project team had engaged a limited set of contractors during the design phase, with the open intention of retaining them to build the telescope facility.

This apparent lack of competition actually helped to reduce overall costs. Larry Simmons made the budget allocation process transparent so that all participants—whether they were from universities, private industry, or NASA—had the same information. Because they were part of the design process, contractors could provide more accurate estimates of how much time and money it would take to build SIRTF; and because they were part of the budget process, it was a little easier to hold them to their financial commitments if the project encountered problems later. This was important, because NASA Administrator Goldin, under pressure from Congress, was taking a hard line on overruns and regaining control over costs at NASA. The success of SIRTF depended on a collective effort by the academic, government, and commercial partners, and Simmons ensured that their common interests were kept in view. To buffer against overruns and to ensure teamwork, Simmons had incorporated some innovative financial incentives into everyone's contracts. As he reported to the SWG in April 1998, "These incentives are contingent in part upon completion of Phase C/D by the entire team with the $450 million cost cap, and upon SIRTF meeting Level 1 requirements 2.5 years after launch, and not solely upon individual team member performance."[13]

It would be unrealistic to think that from here on in, the sailing would be smooth. Although the review boards had strongly recommended that SIRTF move into Phase C, ongoing reviews, similar to the earlier NAR and PDR panels, had identified a few potential problems.[14] The top two were schedule and software. This turned

---

13. SIRTF SWG Meeting Minutes, 21–23 April 1998.

14. SIRTF SWG Meeting Minutes, 11–12 November 1998.

out to be an astute assessment. As the construction of SIRTF got under way, these two issues were the main source of unanticipated costs and frustration.

The schedule was aggressive in part because if SIRTF launched in 2001, all four Great Observatories—each of which had a limited lifespan—could be operating at the same time, realizing Charlie Pellerin's vision. Another upside to an aggressive schedule was that it encouraged everyone to work efficiently. The downside was that if something went wrong, there was minimal time available for troubleshooting problems and testing solutions.

The Hardest Problems

While the software and schedule would provide the team with some difficult moments, neither was the toughest problem, according to Charles Lawrence, SIRTF's new deputy project scientist, who has remarked that the greatest challenge in an innovative project is "figuring out what to worry about, and when to stop worrying about it."[15] In SIRTF's case this principle was illustrated by two problems that lacked a clear resolution: the overpressurization of the dewar and the delamination of one of the spectrographic filters.

In an innovative project, there will almost always be problems that are hard to resolve. First, it is difficult to make decisions when dealing with inherently novel situations. SIRTF was a one-of-a-kind, state-of-the-art telescope facility; what little precedent there was—in sensors, orbits, and cryogenics—might not be relevant. Second, the system was so complex that fixing one problem risked causing new ones. Better the devil you know than the one you don't. Third, the data used to arrive at a decision are often themselves novel and ambiguous. (Recall how the infrared data that Jim Houck and Martin Harwit obtained in the 1960s using rockets was erroneously dismissed as instrument error by other experts.) And fourth, it's often difficult to tell whether or not a solution really addresses the problem, because tests can't always be devised to reveal the impact on all of the affected elements. "We live in a world of room temperature," says Tim Kelly, cryogenic expert at Ball Aerospace. "Our engineering has been developed over 4,000 years at room temperature. Our intuition is at room temperature. When we move off of that, either up or down, we're in unknown territory and intuition doesn't work."[16] The only truly sufficient test is to see how the telescope facility operates in space, but by then it's no longer a test.

Under Pressure

At the end of December 2000, SIRTF almost blew up at Ball Aerospace—literally. Just about everyone who has ever worked with dewars—including most members of the SWG—has blown one up at one time or another.[17] Ball had assembled the flight-ready instruments and was testing their performance under cryogenic conditions by placing them inside the dewar in which the instruments were to fly, along with several hundred liters of supercooled helium under high pressure. At that stage, it's potentially a bomb,[18] and a small misstep can result in a very large problem.

Ball was responsible for almost everything related to the science mission, while Lockheed

---

15. Charles Lawrence, in conversation with the author, at Jet Propulsion Lab, Pasadena, CA, 29 April 2005.
16. Timothy J. Kelly, interview with author, Boulder, CO, 20 March 2009.
17. Houck interview, 25 May 2009.
18. The helium is not explosive, per se. Rather, it is the tendency for liquid helium to rapidly expand as it warms and turns to gas, and this rapid expansion will cause the container (in this case the dewar) to burst.

developed the spacecraft and the flight software. Ball had built two of the three science instruments (NASA Goddard built the third), and the cryogenic dewar, and it had recently completed the assembly of the MIC in which the three scientific instruments were integrated. The MIC was then put into the dewar, which was filled with liquid helium, and Ball ran tests to see how well the instruments operated together at a few kelvins. The tests had gone very well. Ball had overseen the systems integration of dewars for other missions such as IRAS. As a leader in cryogenic technology, Ball had met its schedule commitments. NASA's contract with Ball included performance incentives in the form of bonus fees for meeting certain deadlines. Ball's portion of the project was four months ahead of schedule at the time of the incident.

Over the holidays, one of Ball's technicians was monitoring the live cryostat. Sensors made it possible for him to do so from his home computer, but on occasion he would come to the Ball offices to top off the slowly evaporating helium in order to keep the instruments at the correct temperature. This was a tricky task. Liquid helium is so cold that it will turn any gas or liquid it touches into a solid, including the air around it. As it happened, an ice plug formed in one of the helium vent lines. Dr. Bill Burmester was walking by, just checking on things at the lab. When he saw the problem that confronted the technician, Burmester told him to call Tim Kelly.[19]

Kelly had been with Ball since the 1970s. He had worked on IRAS and knew as much as anyone about cryogenic systems. He had risen to become a senior project leader and was known for putting out fires. (In his case, this description was more than a figure of speech—Kelly was also a volunteer fire fighter in Colorado.) When he heard about the problem with SIRTF's cryostat, he acted immediately: "I organized the communication that had to go out...and made sure things were safe, as best I could," Kelly remembers. "I called the president of the company [Don Vanlandingham], called NASA [Bill Irace, deputy project manager; and Dave Gallagher, who had succeeded Larry Simmons as project manager], told them what's going on..., made sure that [our guy] had somebody over there ASAP to help him—Quality and Mission Assurance.... I got the program manager who was running the program at the time involved." Kelly defused the emergency and turned it over to the team to resolve the ice plug. He continues:

> Now we're into January [2001], about a week or ten days later.... A [different] technician apparently notices the temperature is rising in the cryostat. He tries to tell people in the meeting, the engineers and the program managers—this is Ball people, now.... They're sitting around trying to figure out what's going on. They've got an ice plug and they've got to get rid of it.... They just ignored him, because they said, "If the temperature's going up, the pressure would be going up." Well, guess what, the pressure *was* going up.[20]

The sensors weren't providing accurate pressure readouts because of the ice plug. By the time the evidence was unambiguous, the pressure was so high that people were no longer just worrying about possible instrument damage but whether the cryostat would explode. According to Kelly, people were "not managing the information"; a series of small missteps had occurred and now they had a bomb on their hands, with all of SIRTF's instruments inside. Kelly describes what happened next:

---

19. Kelly interview, 20 March 2009.

20. Kelly interview, 20 March 2009.

So I was called to come back, and the very first job I was given was to work with the team and put that presentation together so that I could spend an hour and a half in front of Ed [Edward J.] Weiler [then Associate Administrator for NASA's Space Science Enterprise] at Headquarters and explain how this went wrong.... We were drinking coffee at NASA Headquarters in the morning, before the meeting, and the Lockheed guy said, "Weiler is going to kill you." Because he's known to be a very non-linear guy.... I'm in this meeting, it's packed. It's a rather small room, but it's packed. Weiler is sitting there in his chair, and I'm giving the presentation. I'm [a few minutes] into it, and [Weiler's] lieutenant says, "Tim, it sounds to me like whatever you could have done wrong, you did do wrong. Is that what you're telling me?" I said, "Yes, that's pretty much it." Well, it's actually a very disarming technique. Now, I didn't mean it as a technique, I really didn't. But honesty always is disarming in those kind of venues. At that point, I was allowed to march through the rest of the presentation, because I wasn't defending myself, so no one was attacking me. They were asking questions. So we had a good meeting. We laid out what went wrong and how we were going to recover. At the end of the meeting, Weiler slumped in his chair. He just slumped. He said, "It sounds like you know what you did wrong and you know how to fix it. Get the hell out of here." I think we all ran. And that was it. We went back and we fixed the cryostat lickety-split, and it's still working.[21]

The problem was finally clear—Ball and the SWG had reconstructed the event and now knew exactly how much overpressurization had occurred, when it started, and how long it lasted. But short of replacing all of the components, the solution could only be a partial one. The SIRTF team would have to rely on tests to determine whether any damage had been done—such as leaks or weakened connectors—that could jeopardize the scientific mission (see Fig. 8.8, p. 128). SWG member Bob Gehrz, working closely with Ball, led a tiger team to test the integrity of the dewar in every way possible.[22]

Coming Apart

"We used up all of our schedule margin in convincing ourselves that the cryostat was OK, and at the same time there was another problem lurking in the weeds," Mike Werner remembers. "There was a problem with one of the filters in the infrared spectrograph, which had shown signs of deteriorating."

The delamination of the spectrograph filter created a situation in which the problem was clear but the solution was not. Like a tiny sheet of plywood, the filter was made of many layers of exotic materials. The first layer was deposited on a crystalline silicate substrate, and the others were deposited one upon the next in a high vacuum environment. The resulting filter was intended to block much of the spectrum, allowing only a relatively narrow range of infrared wavelengths to reach the spectrograph module. Houck had found evidence that the filter had begun to delaminate, which means that some of the layers had separated. Houck's team had delivered a pair

---

21. Kelly interview, 20 March 2009.

22. Kelly interview, 20 March 2009; also see John P. Schwenker et al., "SIRTF-CTA optical performance test," *SPIE Proceedings 4850* (*IR Space Telescopes and Instruments*), ed. John C. Mather, (Bellingham, WA: SPIE, 2003), pp. 304–317.

FIGURE 8.8. The instrument cryostat (with a mass model telescope on top), being prepared for shake testing.

of these filters to Ball and kept a spare set.[23] When technicians placed the detector arrays into the IRS, one of the chips—slightly thicker than the rest—came under greater pressure as the mount was tightened. As a result of this tension, the chip flexed just enough to weaken some of the glued bonds, causing the layers to delaminate at one end. In plywood, this would result in splinters. In a silicon detector, it interrupts the signal. The filter is a multi-coated optical surface—layer upon layer—and the coatings had separated from the substrate, leaving only gold pads holding it in place.[24] The chief concern was that the filter would further delaminate during launch or in space—due to mechanical shock, vibration, and/or thermal cycles—and in the worst case, little particles would break off and float around in the telescope, distorting the data.

The straightforward solution would seem to be to replace the filter. What made the delamination issue so challenging is that it was not discovered until after the detector array had been installed into the MIC. The MIC was by this time sealed and sitting in a dewar of liquid helium. Disassembling the MIC now to fix an existing problem would probably cause new problems. For example, when the three instruments were installed in the MIC the electrical connections which brought the signal from the instruments to the outside were made by hooking together very delicate cables consisting of very fine wires soldered to a miniature connector. The cables were kept as fine as possible in order to reduce the heat that they carried from the warm exterior to the cold interior of SIRTF; as a result they were very delicate. During disassembly or reassembly of the MIC, a wire could detach from the connector, leading to a loss of signal or even a short circuit. In fact, according to Houck, as the MIC had cooled, "some of the connectors in the MIPS instrument [had] failed. There was no question—it hadn't happened when the telescope was warm. The outside of the telescope was warm, but when the outside of the telescope got cold, that's when there was this problem."[25] Given the difficulties of assembling the MIC, getting all three instruments working together was a huge accomplishment, even if one of Houck's spectrographic filters was partially defective. Opening the MIC now was not without risk and could cause both the schedule and the budget to slip by at least a half year.

"Houck...was rightly very concerned that the filter might delaminate," Werner said, "and his instrument wouldn't work well, and we would disappoint our user community, and he would get a bad name, basically, which is a very reasonable thing for him to [have been] concerned about. At first he said, 'I'll take care of this myself,' but it was going on at the same time as this problem with the cryostat, so it didn't get as much visibility. But after we were done convincing ourselves that the cryostat was OK, then we had this potentially delaminating instrument component to worry about."[26] Although the delamination was a potentially serious problem, its loss might affect only one of four modules of the spectrograph; the other three would be unaffected if the fragments did not spread throughout the MIC.

Opening the cryostat had high risks, but it was not without precedent. The IRAS cryostat had been opened in order to fix several shorts. Although this caused a schedule delay, it also

---

23. Ball Aerospace's Optical Coating Laboratory and Boeing manufactured the chips for the IRS instrument's infrared detector array. For more detail, see J. R. Houck et al., "IRS: the Spectrograph on SIRTF; Its Fabrication and Testing," *SPIE Proceedings 4131 (Infrared Spaceborne Remote Sensing VIII)*, ed. M. Strojnik and B. F. Andresen (Bellingham, WA: SPIE, 2000), pp. 70–77.

24. SIRTF SWG Meeting Minutes, 15–16 August 2001.

25. Houck interview, 25 May 2009.

26. Werner interview (Session I), 15 December 2008.

provided an opportunity to correct a few other minor problems. IRAS later performed extremely well in space and must be counted a success even though it cost twice as much as budgeted and took two years longer than planned.[27]

The team was searching for any kind of data or tests that might help them characterize the extent of the delamination problem, in the hopes of arriving at the right solution. Of the four vendors for this type of filter, three had gotten out of the business during the past decade. "We were still waffling around," Werner said. "We went into a shake test which was to simulate the launch, and the filters survived the shake test [in which SIRTF was placed in a simulator that replicated the mechanical forces at launch]. So it was still suspect, but it was no worse after the shake test than it was before the shake test."[28] Presentations were made to Headquarters by Houck and Dave Gallagher, but ultimately a decision needed to be made. It would have to be made by Gallagher, who had taken over as project manager in 1999, when Simmons was asked to set up and lead a new Astronomy and Physics Directorate at JPL.[29] Simmons still had some involvement in SIRTF, but Gallagher, who had been Simmons's deputy, now had top responsibility.

To share opinions and insights, members of the SWG and the project office met over dinner at a restaurant near Caltech. Houck was strongly in favor of opening up the cryostat and replacing the filter. So was Irace, who had been involved in the opening of IRAS during the integration and test phase before launch.[30] To characterize the effects of bits of the filter coming loose in the telescope, Werner noted that chips of paint had affected the NICMOS instrument on Hubble. This problem had been analyzed and documented by the Hubble scientists; however, Irace, who is a tenacious troubleshooter and accomplished systems engineer, felt that those findings were not good enough to use as reliable engineering data. In a rare display of anger, Werner was upset with Irace for dismissing one of the few data points they had.[31] Ironically, Irace rejected data that supported his case to open the dewar, while Werner fought to include data that undermined his case to leave the dewar sealed.

At an SWG meeting after the shake test, there were several who disagreed with Houck and Irace and felt the risks of opening up the dewar were too high. "I thought as chairman of the Science Working Group, I should speak up first to say what I thought," Werner recalls. "This was very difficult for me, because I had known Jim [Houck] for 35 years at least, at that point.... Tom Soifer had been his first graduate student, Tom Roellig had been his student.... I was at [graduate school] with Jim back before either of us had a Ph.D."[32] Soifer, Werner, Frank Low, and Marcia Rieke all voted against opening the cryostat. George Rieke was also against opening it, even though doing so would have given him an opportunity to fix the dead wiring on his

---

27. Martin, telephone interview, 27 March 2009.

28. Werner interview (Session II), 19 January 2009.

29. SIRTF SWG Meeting Minutes, 19–21 October 1999. Larry L. Simmons was officially promoted to head a new Astronomy and Physics Directorate in May 2001, per JPL press release of 2 May 2001 (online at *http://www.jpl.nasa.gov/news/releases/2001/2001reorg.html* (accessed 30 August 2016).

30. IRAS Explanatory Supplement XIII, Contributors to IRAS, available at *http://irsa.ipac.caltech.edu/IRASdocs/exp.sup/ch13/A.html* (accessed 30 August 2016).

31. G. H. Rieke, *The Last of the Great Observatories: Spitzer and the Era of Faster, Better, Cheaper at NASA* (Tucson: The University of Arizona Press, 2006); also Werner interview (Session I), 15 December 2008.

32. Werner interview (Session II), 19 January 2009.

instrument.³³ "I'm not sure if I made a recommendation formally to [Gallagher]," Werner said, "but if I did, I would have said, 'This is what the Science Working Group assessment was, but in my position as project scientist, I think you shouldn't change it.' They are advisory to me, I'm advisory to him."³⁴

Gallagher took their advice into consideration. Even if they fixed the filter, there was a possibility that new problems would be introduced. No matter what he decided, he would have some unhappy scientists. "I tried to do it in the most collegial way," Gallagher said, "... but ultimately I decided that we shouldn't open everything up. It was a huge schedule hit to open everything up, probably six to nine months."³⁵ The outcome of his decision wouldn't be known until SIRTF was in orbit.³⁶ Fortunately, the filter would survive launch and that module of the IRS operated satisfactorily throughout the cryogenic mission with no obvious further delamination of the filter.

The delamination problem was one of the most stressful events for the team during the entire project. Houck would have preferred a different outcome and wasn't happy with the process at times, but he abided by Gallagher's decision.³⁷ As George Rieke recalls:

> We really discussed these things in a very open way, rather than [making] a unilateral decision, or feeling that there were political games being played, or anything like that. I think what made it work smoothly was just the tone that Mike [Werner] and Dave [Gallagher]—and, obviously, others, but Mike and Dave in particular—set at that stage. [At the dinner] we sat around, and we discussed this—just "Here's a problem, what are we going to do about it?" There was no matter of rank, there were no raised voices, just everybody puzzling over what the best solution was. In the end, we took a vote. Jim Houck lost the vote, and he then sort of ridiculed us gently. He drew little symbols and passed them around, and people who had said things he thought weren't very courageous got a waffle symbol.... Just a little drawing of a waffle. But that was it. It illustrates a mode of problem-solving that's very hard to achieve but very effective. What's really effective about it is that when you solve the problem, you haven't created wounds that make the next problem even harder to solve.³⁸

The Software Challenge

While the problems with the dewar and the delamination were surprising, other problems were less so. As the Phase B review board had noted, the software was likely to be a source of trouble. Although SIRTF's software would be partly based on prior missions, some fundamentally new activities would require mostly custom software. First, SIRTF had three instruments that operated together and made systems integration more demanding, whereas instruments on

---

33. According to Werner, "The MIPS team lost part of its 160-micron array to a wiring failure before launch and half of its 70-micron array after launch, probably to the same cause" (personal communication with author, 8 November 2010). Despite the wiring problem, Rieke's instrument still demonstrated excellent data-acquisition capabilities.
34. Werner interview (Session II), 19 January 2009.
35. David B. Gallagher, interview with author, Pasadena, CA, 3 March 2009.
36. In a later exchange Werner added, "Fortunately, [the filter] survived launch and stood us in good stead throughout the flight. Whatever degradation there was has been dealt with in the data processing and analysis procedures" (personal communication with author, 8 November 2010).
37. Houck interview, 25 May 2009.
38. Rieke interview, 9 June 2009.

previous cryogenic missions had operated independently. Second, the pointing control system (PCS) that aimed the telescope toward observational targets needed to operate with far more precision than its counterparts had on prior missions. Third, SIRTF needed to be more adaptable in detecting and recovering from faults while in orbit. Ordinarily, when a spacecraft encounters a problem, it shuts down and goes into safe mode, powering down and awaiting instructions from ground crews. It can take a week to restore science operations from this safe state. SIRTF, whose operation was already limited by its cryogen stores, couldn't afford to take a week to recover each time it ran into a problem. Instead, software was needed for a new "standby" mode, in which SIRTF operated autonomously when dealing with a problem, shutting down a limited portion of the facility, thereby enabling ground crews to restart operations within a day or two. Fourth, SIRTF needed software to handle the unprecedented rates of data collection; it was the first mission to make use of the new 2.2-Mbps download capability of the Deep Space Network. Fifth, even existing spacecraft flight software needed to be redesigned to account for SIRTF's novel Earth-trailing orbital dynamics. All of these innovations made developing the software far more complex than it had been for prior missions.[39]

Responsibility for developing the flight software and the spacecraft went to Lockheed Martin Space Systems Company (LMSSC). In 1995, Lockheed and Martin-Marietta merged. This brought together the space systems group in Sunnyvale (formerly Lockheed) and the flight-operations group in Denver (formerly Martin-Marietta). When the Phase B proposals were submitted in 1996, the merger provided a compelling reason to have the combined resources and experiences of Lockheed Martin brought to bear on the development of SIRTF. While Sunnyvale would have overall responsibility, it was believed that some of the flight software previously built by the Denver team could be repurposed and integrated into the software being developed in Sunnyvale. One person who thought so was Frank Martin, who had preceded Charlie Pellerin as NASA's Director of Astrophysics and laid the foundation for the Great Observatories. Martin was an executive with Lockheed Martin at the time of the proposal. After they had won the contract and software development had begun, he recalls, "we stopped and looked at it carefully, [and] it really didn't make as much sense. This is one of those cases where it's a better marketing story than it is an implementation story. The Lockheed team, after I left [Lockheed Martin], struggled quite a bit in getting the software straightened out."[40]

The software requirements were more complex than those in anyone's prior experience, and the challenges of the corporate merger were underappreciated at the time. "I think [Lockheed Martin] erroneously assumed that there was a lot of stuff they [could] reuse," Mike Werner said. "They didn't realize the complexity of [SIRTF's] fault protection, which relies on autonomy, because [SIRTF] is not in constant contact with the Earth, so if something goes wrong, it has to be able to sense it and correct it, and that's called autonomy." For every software component—whether new or repurposed—there had to be a fault-protection complement. "I think the complexity of that totally caught them off guard," Werner said.[41] Even JPL's Johnny Kwok, who proposed SIRTF's solar Earth-trailing orbit and has

---

39. Patricia Lock, "SIRTF—Inheritance, Adaptation, and Advancement," in *Proceedings of AIAA SpaceOps Conference 2000*, (June 2000), available at *http://hdl.handle.net/2014/14468* (accessed 30 August 2016).

40. Martin telephone interview, 27 March 2009.

41. Werner interview (Session I), 15 December 2008.

developed software for several successful NASA missions, expressed frustration that he couldn't do more: "I could contribute [to] the communication system, the pointing-control system, and some of the functionality on the spacecraft," Kwok said, "[but] I don't know what to do with the flight software. It's like spaghetti, and I don't know how to get it unraveled."[42]

Lockheed Martin tried to solve the problem by adding programmers. The software continued to fall behind schedule. Bill Irace, who was overseeing the systems engineering for SIRTF, was spending two or three days a week at Lockheed's Sunnyvale facility; however, he, too, was unable to tame the problem. "[We didn't realize quickly enough] that Lockheed Martin didn't know how to write the software," Irace remembers.[43] The "faster, better, cheaper" policies of Dan Goldin had strongly encouraged the repurposing of software as a way to save money in development and testing. But in practice, those savings are very difficult to achieve unless the team comes with the software, because there is always some implicit knowledge that resides in the staff and is not captured in the code or documentation.[44] Knitting together a team was hindered by the recent merger of Lockheed Martin. "They were totally different [organizational] cultures," said Gallagher. "That created a lot of problems in getting a team together."[45]

Compounding the issue was a software labor shortage—it was the late 1990s and programmers were shunning established firms in favor of Internet ventures. "I remember they had job fairs right outside the Lockheed Martin gates in Sunnyvale. It was very hard to keep people on the team," Gallagher said.[46] As a result, the software team at Sunnyvale was fairly new. Kwok says, "I remember they were sending people to some of the basic real-time software control classes. I said, 'Wait a second! You're building my telescope and you just go take a class on how to write real-time control software?'"[47] Sunnyvale had project oversight but was relying on the team in Denver to produce some of the electronic parts and share their considerable flight-software experience. "But they had not learned how to work together," Kwok said.[48] It probably did not help that the Sunnyvale and Denver operations had both wanted to bid on SIRTF. Lockheed Martin management had picked Sunnyvale to lead and expected Denver to simply help them.[49] But in practice this turned out to be problematic, as Denver focused on its own contracts, many of which were for NASA's high-profile Mars missions. Lockheed Martin changed project

---

42. Kwok interview, 25 March 2009.

43. William R. Irace, interview with author, Pasadena, CA, 18 February 2009.

44. Robert K. Wilson and Charles P. Scott, "The Road to Launch and Operations of the Spitzer Space Telescope," paper presented at the SpaceOps Conference, Rome, Italy, 16 June 2006; available at *https://ntrs.nasa.gov/search.jsp?R=20080022114* (accessed 30 August 2016); also see James L. Fanson, Giovanni G. Fazio, James R. Houck, Tim Kelly, George H. Rieke, Domenick J. Tenerelli, and Milt Whitten, "Space Infrared Telescope Facility (SIRTF)," Proceedings of the Society of Photo-Optical Instrumentation Engineers (SPIE), Space Telescopes and Instruments V, no. 3356 (August 28, 1998), pp. 478–491.

45. Gallagher interview, 3 March 2009.

46. Gallagher interview, 3 March 2009.

47. Kwok interview, 25 March 2009.

48. Kwok interview, 25 March 2009.

49. Simmons interview (Session II), 9 March 2009.

managers several times, but this had little effect on SIRTF's software-development rate.

Eventually, the SIRTF project team enlisted the help of Headquarters. In a progress report to the SWG, Irace mentioned that "the status of SIRTF flight software is a major concern at Headquarters, with Administrator Goldin inquiring about this topic. Weiler [Associate Administrator for Space Science Enterprise] has discussed this issue with the CEO of Lockheed Martin."[50] Headquarters had already delayed SIRTF's launch, for reasons that had as much to do with the software problems as with a shortage of launch vehicles. By May 2002, the software was back on track and launch was scheduled for 2003.[51] Lockheed Martin had appointed John Straetker as its SIRTF program manager and Nick Vadlamudi as the observatory system-engineering manager, and by all accounts they turned things around. Straetker "saw the observatory from completing the test and integration at Lockheed through to launch at the Cape. It made a tremendous difference," Werner said. "The spacecraft has worked very, very well. What that may mean is that the lower-level engineers who were working on the project were doing well, but someplace in the middle there was a big disconnect."[52] Simmons concurred: "[T]o the credit of the people working on the job day by day, they sort of fought their own management to make it successful."[53] In the end, the software worked very well, but its cost went from $15 million to $70 million and used up more than half of the budget reserve that Larry Simmons had set aside.

If the software was complicated for Lockheed Martin to build, it would also be difficult to operate in space. Mission operations had been planned on a shoestring, before the software issues were fully understood. The reserve fund might have helped, but the software problems had drawn it down. There was separate funding for SIRTF's Science Center operations, where the data would be received and processed. The SSC was part of the Infrared Processing Analysis Center, which received funding directly from Headquarters to handle several major infrared datasets. SIRTF's mission operations were the responsibility of JPL, but the money was insufficient to properly staff and train a crew on novel software. "For this and a variety of other reasons, we didn't really have a very robust operations plan…, the nuts and bolts of commanding a spacecraft," Werner said. "We actually failed the Critical Design Review for the mission-operations system. At that point, Dave Gallagher went around JPL and said, 'We need the very best mission-operations guy you've got. Who is that?' They said, 'It's Bob [Robert K.] Wilson.' Bob came in and brought in some really good people and straightened things out pretty quickly."[54] Wilson was made the SIRTF mission operations system manager, and Johnny Kwok became the mission system engineer. Keyur Patel, an expert on autonomous software, was made the flight-engineering office manager, and Fernando Tolivar joined the team as flight system engineer. Wilson had 18 months to get everything in order.[55] And he did it. "The big decision that [Wilson] made that was really positive was to enlist Lockheed Martin, in Denver, in the spacecraft operations," Werner said. "Lockheed, Sunnyvale, which is where Spitzer had been built, sort of had the

---

50. SIRTF SWG Meeting Minutes, 15–16 August 2001.
51. SIRTF SWG Meeting Minutes, 10–11 May 2002.
52. Werner interview (Session II), 19 January 2009.
53. Simmons interview (Session I), 18 February 2009.
54. Werner interview (Session II), 19 January 2009.
55. Wilson and Scott, "The Road to Launch," 2006.

right of first refusal, but they really didn't want to do it.... Whereas at [Denver,] there are groups of people whose job it is to operate spacecraft. They operate a lot of [JPL's] Mars spacecraft and were familiar with the spacecraft avionics and electronics.... They were a very logical choice to do the operations. [The Denver team] has done very well by us."[56]

Launch

The Space Infrared Telescope Facility launched from Cape Canaveral Air Force Station in Florida on Monday, 25 August 2003, at 1:35 a.m. Eastern Daylight Time (see Fig. 8.9). SIRTF entered its solar Earth-trailing orbit and opened its lens to the sky, receiving first light on 1 September.[57] It began collecting data on 1 December.[58] The main mission lasted twice as long as expected. The cryogen finally ran out on 15

FIGURE 8.9. Image of the August 2003 launch at Cape Canaveral Air Force Station, Florida, of the Delta rocket that carried Spitzer into trailing-Earth orbit (NASA Kennedy Space Center).

---

56. Werner interview (Session II), 19 January 2009.

57. SIRTF Science Center Press Release #146 (9 March 2003), available at *http://www.spitzer.caltech.edu/news/146-ssc2003-03-Space-Infrared-Telescope-Facility-Mission-Status* (accessed 30 August 2016).

58. The date of September 1 is drawn from the caption of an image transmitted during Spitzer's first power-up sequence, or "aliveness test," in space. Two versions of this image (with and without annotation) can be viewed online at *http://photojournal.jpl.nasa.gov/catalog/PIA04724* (accessed 30 August 2016). For more details, see John W. Miles et al., "Execution of the Spitzer In-Orbit Checkout and Science Verification Plan," a paper presented at the SPIE Astronomical Telescopes and Instrumentation Conference, Glasgow, Scotland, 21 June 2004.

May 2009.[59] SIRTF, renamed the Spitzer Space Telescope in honor of Lyman Spitzer, the pioneering promoter of space-borne astronomy, will continue to send back data with the two short-wavelength detectors on IRAC that can operate at warmer temperatures (34.5 K).[60] By 2016, Spitzer had drifted more than 100 million miles from Earth.[61] Eventually, however, data collection will end as the telescope's signals become too faint for two-way communication on the current data network. Around the year 2068, Spitzer will once again be within signaling reach when its orbit comes full circle. What remains are the data Spitzer has gathered, which have already led to discoveries of exoplanets and observations of the farthest reaches of space.[62] In the data still to come, there are bound to be surprises. The harvest from SIRTF will sustain future generations of infrared astronomers and those who want to understand the origins of our universe.

---

59. SIRTF Science Center Press Release #436 (15 May 2009), available at *http://www.spitzer.caltech.edu/news/436-ssc2009-12-NASA-s-Spitzer-Begins-Warm-Mission* (accessed 30 August 2016).

60. Spitzer Fact Sheet.

61. SIRTF is drifting away from Earth at a rate of 0.1 Astronomical Units (AU) per year. In 10 years, it will have drifted over 90 million miles (per Spitzer Fact Sheet). Its current location can be found at *http://www.spitzer.caltech.edu/mission/where_is_spitzer* (accessed 30 August 2016).

62. For more on the key scientific findings made possible by the Spitzer telescope, please visit *http://irsa.ipac.caltech.edu/data/SPITZER/docs/spitzermission/missionoverview/papers/* and *http://irsa.ipac.caltech.edu/data/SPITZER/docs/spitzermission/observingprograms/legacy/* (accessed 30 August 2016). For a summary of initial findings, please see Michael Werner, "Spitzer: The First 30 Months," *Astronomy & Geophysics* 47, no. 6 (2006): 6.11–6.16; and Michael Werner et al., "First Fruits of the Spitzer Space Telescope: Galactic and Solar System Studies," *Annual Review of Astronomy and Astrophysics* 44 (September 2006): 269–321.

CHAPTER 9

Making the Invisible Visible

Spitzer's success is due in part to its many technological innovations, such as its sensors, its Earth-trailing orbit, and its use of radiative cooling. But it is also due to managerial innovations that can help shape collaboration and sustain a project. Three behaviors fostered the project's success. Based on themes that participants repeatedly mentioned, the three factors were nurturing relationships, challenging boundaries, and establishing interfaces. While these set the management of the Spitzer project apart from others, they are still general enough to be applicable to other projects.

Management Lessons

How NASA manages projects has already been the subject of several books, including one by Charlie Pellerin.[1] The technology that makes Spitzer unique is truly remarkable, and all of the SWG members have written detailed descriptions of the instruments and the facility. Over seven thousand articles have been published on the scientific results from Spitzer. There are even two personal histories of the project, by Mike Werner (2006) and George Rieke (2006). What all of these narratives skim over, however, are the managerial skills that enabled it to endure for decades, to transform itself from orbit to orbit, and ultimately to secure the commitment of shifting constituencies. Management was not the primary focus for most of the people working on what was then known as SIRTF. Because bad management tends to stand out, while good management fades into the background, the nearly invisible ways in which this project was managed are worth a closer look.

In more than 35 interviews, most of them conducted in 2008 and 2009, participants repeatedly mentioned the management contributions of two people: Mike Werner and Larry Simmons. Werner has been the project scientist since 1983. Whereas other members of the SWG worked on other missions in addition to SIRTF, this project was Werner's full-time job. "I have to give Mike...particular credit for keeping the project alive," Simmons says.[2] In turn, Werner credits Simmons with keeping the team aligned and making SIRTF happen. Simmons was a relative newcomer, joining as project manager at the

---

1. Charles J. Pellerin, *How NASA Builds Teams: Mission Critical Soft Skills for Scientists, Engineers, and Project Teams* (Hoboken, NJ: Wiley, 2009).
2. Simmons interview (Session I), 18 February 2009.

end of 1993, but he possessed deep institutional knowledge about how to get things done at JPL.

"The project scientist and the project manager are partners in assuring the scientific success of a mission," notes Werner.[3] Nevertheless, both Werner and Simmons would quickly avow that much of the credit for SIRTF's success belongs to other people. This collaborative spirit is partly what made them so effective. Others also played key roles: Fred Witteborn and Nancy Boggess got SIRTF started; Martin Harwit gave it a storyline; Charlie Pellerin, Dan Weedman, and Larry Caroff worked the politics and budgets at Headquarters; Frank Low and Johnny Kwok came up with technical breakthroughs; and Marcia Rieke and Bob Gehrz worked to get Congress and the scientific community lined up behind SIRTF. Dozens more who are not mentioned by name supplied valuable assistance, but those identified here stand out because their actions were striking or tied to specific, visible events. While their contributions were necessary, they were not sufficient. It was Werner and Simmons who provided the cement that held everything together.

Although Werner and Simmons had different roles, they had the same goal: to get SIRTF into orbit. No manager can force people to get along, but he or she can establish structures and norms that shape team behaviors, such as courtesy and collaboration. SIRTF's managers successfully changed the relational aspects of teaming by holding retreats and team-building workshops, rotating meetings to different locations, colocating teams, and establishing contracts that rewarded team outcomes.

Simmons and Werner pioneered these efforts, but they were emulated by other senior leaders for SIRTF, such as Dave Gallagher and Bill Irace at JPL, Tim Kelly at Ball Aerospace, and John Straetker at Lockheed Martin. Because direct testimony can best illustrate how these managerial techniques were applied in the case of SIRTF, the following sections contain interview excerpts that address each technique in turn.

A. NURTURE RELATIONSHIPS

One mechanism used to organize the project activity was to forge interpersonal relationships. Strong relationships can help to hold a project like SIRTF together, especially during crises or long delays. But most relationships do not exist before the project. They are established as people are brought together to work on it, and those relationships need to be developed, regularly maintained, and occasionally repaired. As the top two people officially responsible for managing SIRTF, Werner and Simmons (and later Gallagher) felt that nurturing interpersonal relationships was one of their most important contributions—even though they sometimes had to defend this activity to their team members and upper management.

1. Spend time in their world. Managers actively developed their own connections across the organization. They thus knew in advance where particular knowledge and information resided, and they were well positioned to detect the first signs of emergent problems.

Dave Gallagher, SIRTF Project Manager (JPL)
"I learned this a long time ago. When I was back at Headquarters, I would just sort of walk through the building. I made a lot of friends and got to know a lot of people. It sounds lame, but I would start on the top floor, and I would basically walk the floor and get to know people and introduce myself to people I didn't know and try to establish some relationships…. It wasn't manipulative in nature, like 'I need to get to know these people because I'm going to want something from them.'

---

3. Werner interview (Session II), 19 January 2009.

I actually enjoyed it. I like to understand organizational dynamics and how things really work and how decisions get made and what motivates people. The best way to understand that is [by] spending time talking to people, so that's what I did."[4]

Mike Werner, SIRTF Project Scientist (JPL)

"A very important part of this kind of a job is to be proactive. I would spend a fair amount of my time talking to the other managers just to make sure I understood what was happening. At one point we had a concern about how we were going to get the telescope focused. Bill Irace and I purposely identified a couple of people we thought would be good to head up that activity—a couple from the science team and one engineering guy who set up this tiger team that worked on focusing the telescope. I participated in those discussions and made sure that we had a well-defined, agreed-upon focus criterion that was argued out in advance, so that we wouldn't get into a big argument about it in the heat of battle. I would take on special topics that were cross-organizational and that otherwise might have fallen through the cracks."[5]

Dave Gallagher, SIRTF Project Manager (JPL)

"I'm very focused on trying to spend a lot of time with people, trying to…understand their world. That requires time and travel. This team's key players…had all been together so long, you know, that I was definitely a little bit of an outsider [in 1999]. There was an element of earning their trust, spending time with them, understanding how we'd gotten here. That was one thing. Keeping our relationship with NASA Headquarters healthy was extremely time-consuming. Lia LaPiana [SIRTF program executive at Headquarters] and I spent a lot of time on the phone and either her coming out here [to JPL] or us going back there [to Headquarters] sort of monthly. I borrowed—and maybe modified a little bit, but mostly just borrowed—this concept from Larry [Simmons] of a monthly management team face-to-face meeting. I think that's very important. We rotated the venue.… We'd go to [the University of] Arizona, we'd go to Cornell, we'd go to Lockheed, Ball. It really helps to meet with people on their turf. Again, that's time-consuming. We ultimately had to go get more money from NASA Headquarters. That process was challenging."[6]

2. Enforce respectfulness. Many SIRTF participants had joined the team as the result of winning a competition, including the PIs, the NASA Centers, and the contractors. Simmons had to find a way to convert the dynamics of competition into those of collaboration. He established a rule that team members would not complain about one another. SIRTF was being developed with limited resources; there were plenty of challenges to address but not enough money to address them all. Every member of the team needed to do his or her part, but in order to meet their own obligations, they often had to rely on others.

Bill Irace, SIRTF Deputy Project Manager (JPL)

"I think if you had to point to one thing that caused SIRTF to be successful, it was that it was a culture that was established at the beginning, which is, you do not fight with each other. You do not badmouth someone else. You do not say, 'He doesn't know what he's doing.' Larry [Simmons] never let that happen anywhere in his presence, including in the [project] office itself.

---

4. Gallagher interview, 3 March 2009.

5. Werner interview (Session I), 15 December 2008.

6. Gallagher interview, 3 March 2009.

People would grouse, and he'd say, 'Do not say that. They're a part of the team; we need to work with them. They're the ones who are going to get the job done. So do not ever do that.' That culture was established from the beginning."[7]

Larry Simmons, SIRTF Project Manager (JPL)
"I've often told my teams over the years, you never call a company, you call a person at the company. If the company answers the phone, hang up, because it's a machine. So you come to realize that even though you're not in the business of dealing with people in a buy/sell relationship so much, the thing that makes the project work or not work are the people. So then you have to sit back and say, well, who are the important people and what are the roles of the various people and how do I deal with them."[8]

3. Create opportunities for interaction. In a project as complex as SIRTF, meetings are a fact of life. While the ostensible goal of a meeting is to gather or disseminate information, Werner and Simmons consciously used meetings as a way to foster human connections. Face-to-face meetings were regularly held at different team members' locations. This gave the whole team (not just the leaders) a chance to learn about one another's worlds. It also distributed the burden of traveling to meetings. Many interviewees identified these monthly meetings as essential to SIRTF's success.

Bill Irace, SIRTF Deputy Project Manager (JPL)
"Larry would require the key managers from all the areas, maybe ten people, to meet face-to-face every month, whether they wanted to or not—whether they thought it was a good idea or not. They would always physically meet somewhere; and we would meet in different places, in different people's plants. We'd go to Colorado, or Sunnyvale, or the launch site, or wherever, but we would always meet. That created a human connection among all these people, which was built not just formally but at dinners, parties on occasion."[9]

Larry Simmons, SIRTF Project Manager (JPL)
"We saw each other every month. I attribute a lot of the project's success to those face-to-face meetings. My meetings were four hours long. We'd meet from 8:00 [a.m.] to 12:00 [p.m.], or something like that.... Everybody would come in the day before ... and they would get together for dinner, and then they would meet for breakfast and we would have the meeting. It's too late to go to the airport to catch the flight home, so they would spend the afternoon talking to each other about stuff, and then they would go home the next day. So there was a lot of opportunity for face time.... It facilitated people interacting with each other, and that was really, I think, the key to SIRTF.... Mike [Werner] was excellent at having regular Science Working Group meetings."[10]

Mike Werner, SIRTF Project Scientist (JPL)
"One of the things that [rotating the venue] did is it made everybody feel like they were involved in everything."[11]

4. Create a place for interaction. Despite the efforts to build connections among SIRTF's team members, the fact remained that they were all members of different organizations, spread over three time zones. Simmons cajoled JPL management into giving him physical space so that the project had

---

7. Irace interview, 18 February 2009.
8. Simmons interview (Session II), 9 March 2009.
9. Irace interview, 18 February 2009.
10. Simmons interview (Session I), 18 February 2009.
11. Werner interview (Session I), 15 December 2008.

a home base for team members, regardless of their organizational affiliations. Simmons was thus able to dissolve organizational boundaries and make people feel that they were part of one organization, the SIRTF project. Sharing an office also created additional opportunities for interaction.

Larry Simmons, SIRTF Project Manager (JPL)
"Facilities are always a problem at JPL; there's never enough space. So I lobbied for some space. It turned out that an old building was being refurbished, and the top floor of the building was about to be completely renovated. It was about the right amount of square footage for our project.... So I got to help define the final layout of that floor because we were going to be [its] first users. I could put my entire team up there. I got people from the line organization to co-locate with us. I'm a big believer in co-location. I don't care who your boss is, if you're working on my project, I'd like you to be kind of close by. So we got enough space so that we could co-locate many of the key people. In co-locating them, we were kind of isolated [from the rest of JPL] because this building is sort of off in the corner. That had its good parts and its bad parts, from some people's point of view. We were no longer right in the middle of the swirling JPL. But we were co-located, so if you wanted to ask the structures guy a question, he was right over there. Many of the team members were right next to each other. They'd go to lunch together.... Some of them would stay and talk after work about what we should do about the thermostat, or something like that.... It helped congeal the team. We were able to create office space for some of our contractors.... They were only there part of the time [and desks would sometimes be empty]. So I had to kind of stand there with a straight face sometimes and people would say to me, 'Well, do you really need that space?' I would say 'Oh, yeah, it's absolutely essential we have that space.' So we got it, and that helped, too, because when the scientists came in, there was a place for them [to] sit and work. When the contractors would come, there was a place for them to work, and they could leave notes to each other. We actually established that as a precedent.... Those are the kind of things we could do because there really weren't solid rules at the time. You could kind of create these rules—that we needed places for these people. Colocating them, having them not think of themselves as working for the company as much as working for SIRTF—all that contributed to the glue that made them a team. It became a very strong team.... When you have enough skill, diversity, and quality of team members that almost anything you can come up with is within their capabilities, you have a strong team."[12]

5. Establish the desired norms. Shared space and frequent interactions are often not enough to bring a team together when its members belong to different organizations. Even in the case of the merger between Lockheed and Martin Marietta, corporations that shared a top management team and stock price, the two were more separate than integrated when it came to SIRTF. With little or no organizational leverage, Simmons had to find a way to bring together team members—from various universities, government labs, and contractors—who were not just from different organizations but from institutions with radically different incentives and rewards. Werner helped people to see how their work on SIRTF contributed to a scientific quest. Simmons held a team-building retreat, which was a highly unusual thing to ask of scientists and engineers, whose training had not focused on skills necessary for collaboration, such as networking and interpersonal communication.

---

12. Simmons interview (Session II), 9 March 2009.

Larry Simmons, SIRTF Project Manager (JPL)

"Once we had Ball and Lockheed as part of the team, the very first thing I did is, I had a three-day retreat up in Oxnard [California], at a place called the Mandalay Bay, and invited each of the companies to send—I don't remember how many people, three or four people. The PIs came and the facility scientists came, and the JPL people came, and we just got to know each other. They thought I was a screwball because I said, 'We're not going to get anything out of this meeting except to find out who each other are. I want to know who's got kids,' and so forth.... I put a lot of effort into forcing the team to function as a team.... I had two people from JPL who were in our training organization who actually facilitated [the retreat]. They came up with some of the stupid stuff we did. A guy who's a tenured professor at a university, you say to him, 'I want you to make this thing out of toilet paper tubes and plastic bottles and stuff.' But they were all having a good time.... You just got to know each other. You got a sense for what their personal issues were as well as their professional issues.... I'm experimenting here... [They] got to know each other well enough that...people actually got to where they were working on SIRTF, instead of working for their company."[13]

Frank Martin, Director of Astrophysics (NASA Headquarters):[14]

"When people are all part of one organization—one NASA Center, or one company—they tend to behave differently when they get in a tough situation; if they're partners with somebody, then who's responsible?... Once you get into the mission, people don't pay attention to what badge you're wearing. You're part of a team. When people work on projects, they tend to identify more strongly with the project than they do with their home organization."[15]

Larry Simmons, SIRTF Project Manager (JPL)

"I never let them talk about the company they worked for; they always had to talk about the fact that they worked on SIRTF and what their job was on SIRTF. They were responsible for the cryogenics, they were responsible for the telescope; whatever their job was on SIRTF, that's who they were. They weren't the guy from Ball, they weren't the guy from Lockheed, they weren't the guy from [JPL's] Division 32. They were the SIRTF guy responsible for data processing, or the SIRTF guy responsible for mission analysis, or whatever their job was. It didn't take too long, frankly, because I would hound them about that. It didn't take too long before they quit introducing themselves to each other as 'I'm from Ball.' They'd call each other [and say]: 'I'm responsible for the cryogenics, and I need this from you.'... There was one case where [a person from Lockheed] needed some tungsten to do something with. 'I need a block of tungsten, but I can't get it for three weeks.' [A person from Ball said] 'I've got some here. I'll send you some.' They FedEx'ed it back and forth.... They didn't have contracts with each other, so they were able to share information, share actual materials, because it all contributed to getting the project done—without having to go through a central focus, which some projects tend to do. Some project managers tend to want to know everything that's going on all the time and have everything approved. I didn't do that."[16]

---

13. Simmons interview (Session I), 18 February 2009.

14. Frank Martin was NASA's Director of Astrophysics from 1979 to 1983, the period during which some of SIRTF's initial studies were commissioned. Later, Martin was a senior executive at Lockheed Martin (1990–2001) when the Phase C contracts for building SIRTF were signed.

15. Martin telephone interview, 27 March 2009.

16. Simmons interview (Sessions I and II), 18 February and 9 March 2009.

B. CHALLENGE BOUNDARIES

The second mechanism used to organize the project activity was a redefinition of the boundaries people took for granted. Challenging boundaries is necessary to make people see beyond them. Job titles, a building block of the organizational chart, are intended to clarify who owns what part of the project. Titles can, however, impede creative problem solving and lead to an "it's-not-my-job" mentality. In addition, job titles and boxes on an organizational chart do not reflect the other identities people have within their profession, company, or division. Simmons and Werner removed the traditional antagonisms—between scientists and engineers, government employees and contractors, research and operations—and helped people to instead identify with the project. Team members were made to feel that they were contributing to something larger—and to understand how their piece completed the puzzle.

1. Do not hoard information. On an innovative project like SIRTF, where the design cannot be well defined or the challenges fully understood at the start, people will quickly add buffers by padding the schedule, the budget, and their contracts. This is a normal response to sparse or ambiguous information. Scientists, by training, tend to seek out more information. The adoption of the Earth-trailing solar orbit suggested by Kwok, a lower-level engineering manager, demonstrates how flows of information can benefit the project. In contrast, many of the participants were accustomed to working on classified contracts, or in highly political environments, where there can be real benefits to damming the flow of information. Simmons and others overcame this tendency by sharing essential information with the entire management team, not just those whose paycheck was signed by JPL. With shared information, the team was better able to reach consensus and, when there were disagreements, to work out compromises or alternative solutions.

Mike Werner, SIRTF Project Scientist (JPL)
"One of the things Simmons did which was very effective was he kept open books, so everybody always knew how much money everybody else was getting. We had a management team that met every month; that was the internal management team of the project. And one of the things that was always on the agenda was potential allocations of our project reserve. So even though he, and later Dave Gallagher, reserved the prerogative to make the final decision, everybody could feel that they'd been consulted and had a say in how these reserve funds were allocated."[17]

Charles Lawrence, SIRTF Deputy Project Scientist (JPL)
"Everybody [on the management team] knew how much money there was available. Everybody knew where it was going.... If somebody was having a problem and they needed more money, they made the request. 'Here's why we need this, here's how much it's going to cost, here's what we're going to do.' So everybody had insight, everybody had knowledge. And along with that comes a confidence that you're not missing something, that nobody's trying to pull a fast one on you. That's a departure from the way a lot of things are managed. A lot of managers see knowledge as power and withholding knowledge as one of the main ways of controlling things. I happen to think that's not very productive. If power is what you're interested in, maybe it's OK. If making things work is what you're interested in, it's not very productive.... Now, another aspect was that the contracts as they were written were cost-plus-incentive contracts, as one of the things allowed by federal procurement regulations. The incentive fee had a component that depended on the overall success of the mission. That means that

---

17. Werner interview (Session I), 15 December 2008.

Lockheed's fee depended on Ball's performance; Ball's and Lockheed's performance depended on the instruments working; and so on. The advantage of that is that you've removed the incentive for people to solve their own problems at the expense of somebody else—or to not help solving somebody else's problem in the optimum way just because it's going to cost you a little bit more money. All right, so it's going to cost you a little bit more money. You say, 'Sure, we're glad to do it. It's going to cost us a little bit more money. Overall it's the best solution. Money flows from one place to another, whatever.' So the incentive is you work together, you [get] the best overall product. You don't maximize your little piece of it."[18]

2. Focus on talent, not titles. There is a tendency among scientists and engineers to focus on their specialty—that is, the science or the engineering. Although there was plenty of hierarchy and specialized expertise in the SIRTF project, the roles were surprisingly fluid. Scientists on SIRTF did not simply design their experiment and then hand it to the engineers to build, as many interviewees noted is typically done on space projects. In the SIRTF project, the scientists (perhaps because they were also instrument builders) often acted like engineers when faced with challenges: redesigning rather than giving up. The engineers also acted like scientists, never forgetting that the goal of the mission was not to build something on schedule and within budget but to bring back infrared data.

Larry Simmons, SIRTF Project Manager (JPL)
"For the system-design team, I had a system architect. His name was Jim Fanson. I had Jim run the system-design team meetings. When you have design teams, they work at different levels and you don't only have one: You have one for the electronics and one for the telescope and so forth. Jim Fanson ran the *Über* system-design team. What they did is, they created much of the documentation that flowed down to the other organizations [i.e., the various subgroups that made up the SIRTF team] to tell them what they had to do. 'This is what the telescope has to do.' 'This is the interface between the telescope and the spacecraft,' and so forth and so on. So I had these key guys, and they all had titles—because everybody likes to have a title. I had a meeting with my team once a week. At one of my once-a-week meetings, the team said, 'Well, we need role statements.' I said, 'Well, our job is to get SIRTF built.' They said, 'Yeah, but I'm responsible for the spacecraft, I need a role statement.' I said, 'Well, you're on the design team. Work with the other guys, get the spacecraft built.' 'Well, we need role statements.' I said, 'OK, write me role statements.' They said, 'When do you want them?' So I said, 'Well, make it two weeks,' or something like that—I don't remember what I said. So they all wrote me role statements. So I sat down and I... talked to them about their role statements, and after I got them from them, I put them in a drawer and never got them out again. They never brought it up again. They were OK with it, but they had to write down what their role was, and they wanted to talk about it. So we did that. Now, at the time, the JPL nominal situation was, you'd write a role statement, they'd sign it, I'd sign it, we'd put it in a book, and then somehow it would become part of the history of the project. 'These were the roles of the key people,' and so forth. Then when you went to give them raises and stuff, you'd drag out the role statements and say, 'This guy is responsible for this and this, and therefore he should get a big raise,' or whatever. But I didn't do that. I just said, 'OK, you want a role statement, write a role statement.' They wrote it, and I put them away. They

---

18. Charles R. Lawrence, interview by author, Pasadena, CA, 19 January 2009.

were happy with that, because they had wanted to write them, and they got to write them, and we were done with it."[19]

Charles Lawrence, SIRTF Deputy Project Scientist (JPL)
"The Science Working Group [was] involved at a very detailed level in what might have been thought of as engineering parts of the project. There's a lot of knowledge on the SWG, a lot of experience, and certainly a deep understanding of what the science goals were and what the effect of various possibilities on the science [might be]. There was also, to go along with it, a lot of experience in doing things for constrained budgets and a recognition of the importance of schedules and budgets. So you had a highly experienced and capable science team with a lot to offer in working out solutions to technical problems. And they did. So when teams were set up to address particular aspects of the mission, you'd put together a team that had people from every part of the project involved with it. And you'd get them all together, and they'd work on it together. It didn't matter whose problem it was in particular."[20]

3. Cultivate ownership. Although the SIRTF team made organizational boundaries somewhat permeable by fostering norms of openness and inclusion, everyone was aware that final authority rests in one person, the project manager. As NASA's representative overseeing the project, the project manager decided how resources would be allocated, what risks were not acceptable, and whether to launch. Simmons (and later Gallagher) never abdicated their responsibility, but they did not rely on their power to manage the project. They relied on other people.

Larry Simmons, SIRTF Project Manager (JPL)
"I expected my team to do the job. I didn't expect them to say, 'This is my part of the job, and I don't care about anybody else.' I expected them to work together and, as a team, get the job done. So I had a mission manager and I had a telescope manager. I had a spacecraft manager. I had a system engineer and all of these various skills. I had them get together and have design-team meetings—which I didn't go to, by the way, because I wanted to let them know it was up to them to come up with the design. I would support what they were doing."[21]

Mike Werner, SIRTF Project Scientist (JPL)
"A car can be built with interchangeable people as well as interchangeable parts. But SIRTF couldn't be, and the very first prototype car probably couldn't be, either. Now, there was, of course, during the project, a well-defined organization, [with] well-defined reporting routes from that organization back to Caltech and to Headquarters. But one of the things I learned from Larry Simmons was that 'It's the people, stupid.' NASA doesn't build anything. JPL doesn't build anything. People within NASA and JPL build things. And the way it works—with JPL, at least—is that a tremendous amount of authority and responsibility is vested in the project manager. The project manager is clearly the single most important person in the organization, and everybody else is sort of a subcontractor.... It was an interesting realization that the instrument teams had both the deliverable role, in which case they reported to the project management, and the science role, in which case they reported to me or discussed it with me. [What] made it a success was that we didn't, in general, interfere with the subcontractors unless they were screwing up. We

---

19. Simmons interview (Session II), 9 March 2009.
20. Lawrence interview, 19 January 2009.
21. Simmons interview (Session II), 9 March 2009.

didn't have this arrogant feeling that often comes in a place like JPL—that regardless of who they are or what they're doing, we know how to do it better, and we're going to go tell them how to do it.... A big part of the success of SIRTF was in getting a great team in and empowering the team and allowing the team members to manage their own part of the job [while] keeping enough of an eye on them so that if things started to go awry, we could help put them back on the rails before they got too far off."[22]

Larry Simmons, SIRTF Project Manager (JPL)
"One of the things about a NASA project is, in the space program, the person who really is supposed to be in charge is the project scientist, because he's the guy who's presumably come up with a proposal to do something, or he's been brought in to lead a team of scientists to do something, and what they're trying to do has to get implemented in a way that they can be successful. So, as a manager who also tends to think of myself as a bit of a scientist, my job isn't to tell people what to do so much as find out what needs to be done and see that that gets done.... I'm not out to become the boss as much as I'm out to find the right people and lead a team that will accomplish what we want to do. Early on in a project like SIRTF, it's not real clear what we want to do. So part of the job of managing this is to get the people who do know to share with the rest of the team, as it exists at that time, what we want to do, and see if we can craft something that everybody will be happy with."[23]

4. Design the project around the people. During the three decades SIRTF was in development, it was managed by many different people—administrators at Headquarters, collaborators at the NASA Centers, and contractors at Ball and Lockheed Martin. In a project of this duration and size, staff turnover was inevitable. The project adapted—intentionally or not—to the talent that was working on it. Werner, Simmons, and Gallagher were keenly aware that it was up to them to ensure the best use of the project's human resources.

Dave Gallagher, SIRTF Project Manager (JPL)
"You do yourself a huge favor by not trying to fix everything yourself, especially on big projects with 300 or 400 people. You try and get the right people in the right place. If the only skill you had was that you could pick people well and put the right person in the right job, that's probably all you'd need."[24]

Tim Kelly, Project Manager (Ball Aerospace)
"One of the things Larry [Simmons] told us right at the get-go was that Lockheed didn't win a bus architecture and we didn't win a telescope architecture. We won the right to participate in the design and architecture of the SIRTF observatory. Therefore we were going to go through, basically, a six-month system-engineering tranche collectively, to figure out how we [would proceed]. This was a very hard pill for everyone to swallow, because, 'Hey, you picked us because we proposed to make this thing fifteen feet tall and paint it green, and now you're saying you don't want it fifteen feet tall and maybe it could be red? We don't understand.' So that was one of the first problems Larry had to deal with—trying to get people in a new mindset of 'How we are going to go forward?' and 'Take those proposals you guys labored over and throw them away, and let's get

---

22. Werner interview (Session I), 15 December 2008.
23. Simmons interview (Session II), 9 March 2009.
24. Gallagher interview, 3 March 2009.

a clean piece of paper and see what's the best way to do this.'"25

Mike Werner, SIRTF Project Scientist (JPL)
"We made the best use of the talents and abilities of the people we had who were, on the science side, uniformly all very highly motivated by the realization of how great the scientific return of Spitzer would be."26

Larry Simmons, SIRTF Project Manager (JPL)
"George Rieke [PI for MIPS] was making these detectors that were going to be good at the long wavelengths, and he was building them in the basement at the University of Arizona.... The way scientific investigation is done is, you find something that allows you to do something that somebody else hasn't done.... Sometimes it's the telescope; usually it's the detector. So NASA and everybody expects that the detectors are always going to be your biggest problem, because it's something no one's done before. With the SIRTF team, they had detectors, and so I was evaluating where we were.... I talked to George, and he showed me his lab, and he's got these guys who have been building these detectors for five, six, or seven years, and [he] said, 'It's going to be really tough getting this work transitioned to industry.' I said, 'Why are we transitioning it to industry?' He said, 'Well, we're not going to be allowed to build flight hardware here at the University of Arizona.' I said, 'Why not?' He said, 'NASA won't let us.' I said, 'I'm NASA, and I'll let you.' He didn't believe me. He didn't think we were going to be able to do that. I said, 'Why should we spend money having someone learn what you can already do?'... When you stop and think about it, it was stupid to think about trying to take something that someone had developed and teach someone else how to do it, while the guys who knew how to do it sat and watched them.... When somebody wanted to do something and it didn't make any sense, we'd say, 'It doesn't make any sense.' When someone wanted to do something that made sense, we'd say, 'OK, let's do it.' We didn't need permission for a lot of things. I think that's where people tend to get stuck in the mud sometimes—they're always looking for someone to give them approval to do something. Frankly, that's the job of a project manager—to say, 'This is the right thing to do; let's do it.'"27

C. ESTABLISH CLEAR INTERFACES

The third mechanism is establishing clear interfaces, the place where one thing comes into contact with another. Interfaces occur between organizations, between levels of management, and between the various engineering specialties within SIRTF. Establishing clear interfaces makes it easier for managers to direct information to where it needs to flow. This is the idea behind the organizational chart. But creating clear interfaces is difficult to do. Like dredging a channel—you need to remain vigilant so that the mud doesn't build up as you are sending valuable cargo up- or downstream; you don't want something critical to get stuck on a sandbar. Likewise, the lines in an organizational chart might appear solid, but if in practice they are crossed or broken, then essential information gets bogged down. Many interviewees noted that when the lines of authority and communication overlapped or became ambiguous, it caused problems for the project.

1. Minimize the interfaces. It is perhaps not surprising that the word *interface* came up so often, as the interviewees were all engineers

---

25. Kelly interview, 20 March 2009.
26. Werner interview (Session II), 19 January 2009.
27. Simmons interview (Session II), 9 March 2009.

and scientists. Good designers know that limiting the interfaces between elements in an engineered system is a sure way to reduce costs and risks while enhancing reliability. For just this reason, SIRTF had very few moving parts. SIRTF's managers understood the need to minimize the number of human interfaces—not informal, social relationships but those points where people are passing responsibility and information back and forth.

Johnny Kwok, SIRTF Mission Manager (JPL)
"There were some technical interfaces, but the majority of the problems arose from just people—not having the right people, not having people work together."[28]

Tim Kelly, Project Manager (Ball Aerospace)
"What's a bad interface? Let's talk about organizations. They don't know how to interact, so you have every level, every person, kind of all coming together and tying themselves in knots with constant phone calls, confusion, bickering, questions about memos, countless meetings to sort out problems and misunderstandings. What we tried to do on SIRTF, what Larry tried to do, was have crisp, clean, and clear communications between organizations on technical and programmatic issues. Bill Irace would use the word *broadband* [to describe how] a lot of information comes across, but only in a controlled, clear way."[29]

Frank Martin, Director of Astrophysics (NASA Headquarters)
"[For example,] the Hubble Space Telescope was a very complicated mission in many ways. It wasn't because the telescope was necessarily hard to do—which it was. It was because there were multiple Centers [and contractors] involved [e.g., Marshall, Goddard, JPL, Lockheed, Perkin-Elmer]. There were a lot of really good people drawn to the project, but what happened was, for a variety of reasons, NASA created some rather complicated interfaces for these people to work in.... I didn't know any better.... I said, 'Creating interfaces is not a bad thing, it's a good thing.'... Only later did I learn, after my second big cost overrun [on IRAS], that these interfaces contribute to your problems."[30]

Charlie Pellerin, Director of Astrophysics (NASA Headquarters)
"The things we didn't understand when we were young and naïve was that every interface costs a ton of money, and inter-Center interfaces are the worst."[31]

Frank Martin, Director of Astrophysics (NASA Headquarters)
"One of the things I've learned in looking back at all these missions...is you've got to keep minimizing the interfaces. You've got to be able to do systems engineering on projects. You've got to be able to keep these things where people can manage them, because these things are hard enough to do just to build them."[32]

2. Manage the interfaces. After the administrators reduce the interfaces, those that remain need to be managed. Someone has to ensure that the parts fit together, information is transferred, problems are addressed, and resources are directed toward project needs. These interfaces are dynamic—teams shuffle, issues arise, and priorities shift—yet the connections across the interfaces must

---

28. Kwok interview (Session II), 25 March 2009.
29. Kelly interview, 20 March 2009.
30. Martin telephone interview, 27 March 2009.
31. Pellerin interview, 19 March 2009.
32. Martin telephone interview, 27 March 2009.

be maintained. The comments from Martin and Kelly reflect how interfaces, good and bad, affect those responsible for oversight and those being overseen.

Frank Martin, Director of Astrophysics (NASA Headquarters)
"If you listen to people talk about Center-to-Center relationships, they always talk about partnering; so no one wants to be in charge, even though you have to have that. So when you've got two Centers doing things, there's that interface. And then when you've got contractors providing things—as with Hubble, where you've got a spacecraft coming in and somebody else provides the telescope, and so forth—somebody has to manage that interface. Somebody has to manage what's going on. Whereas if you just have a project…where all the [associate] contractors reported through the prime, [you could hold] somebody responsible and accountable for managing all that stuff. You weren't doing it yourself, and you were using your talents, your system engineers, and your managers to provide insight and oversight of what was happening, as opposed to actually having to do it [yourself] and no one else looking over *your* shoulder. Every time we had a good success and pretty much stayed on cost and schedule, we had prime contractors and one NASA Center.… When we got in trouble was when we let ourselves drift into this business and its mindset—that we'll just get the best and the brightest, and we'll pick from this Center and that Center, and we'll create these management things, and good people will do good things. It doesn't work that way."[33]

Tim Kelly, Project Manager (Ball Aerospace)
"I was a part of the management team and my focus, my job, unlike the other people on my team, was to ensure that I created a good interface to everybody else. That my opposite numbers at Lockheed, at the different universities, at JPL had what they needed and that we as a team looked unassailable. I didn't want to let other people down. We had a great environment where we never really poked at each other in public, because when any one part was hurt, the rest of us looked like idiots, too. We tried to work that way. And I tried to make sure communication was good. I would spend a lot of time with Bill trying to get into his head."[34]

Frank Martin, Director of Astrophysics (NASA Headquarters)
"The difficulty with these things is that when it gets Center-to-Center, or when it gets company-to-company,…your loyalties are outside. People tend to go native, and when trouble happens, it's 'us against them' as opposed to 'we're all in this together.'"[35]

Tim Kelly, Project Manager (Ball Aerospace)
"Usually you learn things around problems. And that's where teams are formed, not around good times. The formative things are the pressures that are put on a team. There were issues with the flat cables that brought signals out from the detectors, across the CTA [cryogenic telescope assembly], and down into the bus. They had very, very small connectors. They were very fragile. The electronic guys wanted them very big. The thermal guys wanted them very small. There were some issues as to whether they were going to work or not. Everybody started to get wound up. You see that all the time on any problem in any organization on anything. As soon as there's a problem, everybody starts winding themselves up, and pretty soon you've got chaos. Larry was very good at keeping people calm and cool. He

---

33. Martin telephone interview, 27 March 2009.
34. Kelly interview, 20 March 2009.
35. Martin telephone interview, 27 March 2009.

said, 'OK, let's get on top of this, let's get on it in a controlled way. Who's responsible for the cable?' 'Well, Ball's responsible for it.' 'OK, let Ball take the lead in solving the problem.' And then he would get Tom Roellig, who's a good physicist, has a good head on his shoulders, who can contribute—an important thing on Larry's part, not just observing, contributing—[Roellig] can represent the science on this and make sure that this is going to go well.… 'Roellig, can you help them?' 'Yes.' 'Ball, can you accept Roellig to come into your meetings and treat him like one of your people?' 'Yes, we can do it.' So that's a good example of a clean interface."[36]

3. Tailor the context. Organizational structures, such as hierarchy and role statements, are created to make it easier to manage complex projects. However, these structures can take on a life of their own, whereby assumptions and virtual boundaries become real. This is problematic in everything but the most routine tasks. Even in SIRTF, where the scientists and engineers exercised a great deal of autonomy, Simmons and other managers took care that the structures did not unnecessarily constrain their behaviors. To foster innovation, structures must serve the people, not the other way around. Simmons sought to tailor (or eliminate) activities that were not useful.

Larry Simmons, SIRTF Project Manager (JPL)
"To make too many boxes and try and stack them next to each other doesn't work nearly as well as getting everybody in one box and hav[ing] them work together. That's what structure can do, if you're not careful. You'll end up having everybody in their own little box. Think about each of us standing in a cardboard box. If we are all in one box, we can move around, we can help each other. If you fall down, I can pick you up. But if we're all in separate boxes and the line between your box and mine is well-defined, then I can't help you very much."[37]

Mike Werner, SIRTF Project Scientist (JPL)
"When I first came to JPL, I thought, 'Well, everything we're doing'—like a preliminary design review or a requirements review—'is something that's been done hundreds of times at JPL. There must be a template we can follow.' But that isn't true. Every project, rightly or wrongly, tends to tailor the processes to its own needs."[38]

Larry Simmons, SIRTF Project Manager (JPL)
"Much of the structure was perception rather than real. [Headquarters] used to require monthly reports from all the projects. People would put in enormous amounts of time writing these twenty-, thirty-, forty-page monthly reports. I said to the guys at Headquarters, 'Do you ever read those?' *No.* I said, 'So if I quit sending them, it won't be a problem.' They said *no* [laughter]. So we didn't do those. It didn't really cause anybody a problem, because it was a perceived requirement. Because we were strapped for cash, I would push back on everybody: Why do you want that? Are you really going to use it? And the things they really needed and really wanted to use, of course, I gave them. But there were many, many things that just were required [on paper but that didn't benefit the project]."[39]

4. Recognize the importance of trust. In complex projects, at least at NASA, there are many levels of oversight. There are dozens of review boards and layers of management that vet the design,

---

36. Kelly interview, 20 March 2009.
37. Simmons interview (Session II), 9 March 2009.
38. Werner interview (Session I), 15 December 2008.
39. Simmons interview (Session II), 9 March 2009.

the science objectives, the operations. And they can (and do) catch errors and anticipate where problems will emerge. However, as Jim Houck puts it, "Every single day, you're closing another lock that you cannot reopen."[40] Making progress requires having courage that you have done what you can and trusting in those who are doing the rest. There's no other choice when you are listening to the countdown, while your life's work sits strapped onto a rocket.

Tim Kelly, Project Manager (Ball Aerospace)
"Interfaces depend upon trust. You have to trust your teammates. I have to trust Lockheed. I have to trust JPL. Often as not, it's not the organization you trust as much as the individuals. When something goes terribly wrong, your first inclination is to be critical. Again, that's where Larry [Simmons] tried to keep people focused: 'This is not about emotions, let's get back to the problem. We know that whatever organization is involved wants this to work as much as you do. So let's just back off of that and get to solv[ing] the problem.'"[41]

Johnny Kwok, SIRTF Mission Manager (JPL)
"People didn't challenge me [on the solar Earth-trailing orbit]. For whatever reason, no one actually asked me, 'Could you double check, triple check, that this will actually work?' It was radical, but I think the team felt that that was the only solution available at the time. But I was always worried: Had I made any mistakes? No, I didn't make any mistakes.... A few months before launch, I was taking a couple of days' vacation.... Larry Simmons called me on my cell phone and he said, 'Johnny, I just remembered, I never asked you. Have you checked your work?' [Laughter.]"[42]

Charlie Pellerin, Director of Astrophysics (NASA Headquarters)
"The performance ultimately is not driven by the science, because you can overcome the science. It's not driven by the technology; with proper planning and work, you can develop the technology. It's driven by the performance of the team, and it's driven by the context they're living in. I would claim that teams that have atmospheres of high mutual respect and expressed appreciation, carefully addressed shared interests, no difficulties across organizational interfaces, etc.—those teams are going to function at 100 percent of what's possible for groups of people to do together."[43]

---

40. Houck interview, 25 May 2009.
41. Kelly interview, 20 March 2009.
42. Kwok interview (Session II), 25 March 2009.
43. Kelly interview, 20 March 2009.

the science objectives, the operations. And they can (and do) catch errors and anticipate where problems will emerge. However, as Jim Houck puts it, "Every single day, you're closing another lock that you cannot reopen."[40] Making progress requires having courage that you have done what you can and trusting in those who are doing the rest. There's no other choice when you are listening to the countdown, while your life's work sits strapped onto a rocket.

Tim Kelly, Project Manager (Ball Aerospace)
"Interfaces depend upon trust. You have to trust your teammates. I have to trust Lockheed. I have to trust JPL. Often as not, it's not the organization you trust as much as the individuals. When something goes terribly wrong, your first inclination is to be critical. Again, that's where Larry [Simmons] tried to keep people focused: 'This is not about emotions, let's get back to the problem. We know that whatever organization is involved wants this to work as much as you do. So let's just back off of that and get to solv[ing] the problem.'"[41]

Johnny Kwok, SIRTF Mission Manager (JPL)
"People didn't challenge me [on the solar Earth-trailing orbit]. For whatever reason, no one actually asked me, 'Could you double check, triple check, that this will actually work?' It was radical, but I think the team felt that that was the only solution available at the time. But I was always worried: Had I made any mistakes? No, I didn't make any mistakes.... A few months before launch, I was taking a couple of days' vacation.... Larry Simmons called me on my cell phone and he said, 'Johnny, I just remembered, I never asked you. Have you checked your work?' [Laughter.]"[42]

Charlie Pellerin, Director of Astrophysics (NASA Headquarters)
"The performance ultimately is not driven by the science, because you can overcome the science. It's not driven by the technology; with proper planning and work, you can develop the technology. It's driven by the performance of the team, and it's driven by the context they're living in. I would claim that teams that have atmospheres of high mutual respect and expressed appreciation, carefully addressed shared interests, no difficulties across organizational interfaces, etc. those teams are going to function at 100 percent of what's possible for groups of people to do together."[43]

---

40. Houck interview, 25 May 2009.
41. Kelly interview, 20 March 2009.
42. Kwok interview (Session II), 25 March 2009.
43. Kelly interview, 20 March 2009.

APPENDIX A

Contributors to SIRTF

Enormous and sustained efforts on the part of hundreds of people and dozens of organizations were needed to imagine, build, and operate SIRTF. The names of all who participated in this endeavor are given here, organized according to institutional affiliation and role in project management or instrument and system construction.[1]

Management

SIRTF Science Working Group: M. Werner, JPL, Project Scientist; C. Lawrence, JPL, Deputy Project Scientist; T. Roellig, NASA Ames, Facility Scientist; G. Fazio, SAO, IRAC Principal Investigator; J. Houck, Cornell, Principal Investigator; G. Rieke, U. Arizona, MIPS Principal Investigator; D. Cruikshank, NASA Ames; R. Gehrz, U. Minnesota; M. Jura, UCLA; F. Low, U. Arizona; M. Rieke, U. Arizona; E. Wright, UCLA.

NASA Headquarters (program management): N. Boggess, L. Caroff, J. Frogel, F. Gillett, J. Hayes, W. Huntress, A. Kinney, L. LaPiana, K. Ledbetter, C. Pellerin, C. Scolise, H. Thronson, D. Weedman, E. Weiler.

NASA Ames Research Center (project management through 1989): W. Brooks, P. Davis, A. Dinger, L. Manning, R. Melugin, J. Murphy, R. Ramos, C. Wiltsee, F. Witteborn, L. Young.

Jet Propulsion Laboratory, California Institute of Technology (project and science management, mission operations): D. Achhnani, A. Agrawal, T. Alfery, K. Anderson, J. Arnett, B. Arroyo, D. Avila, W. Barboza, M. Bareh, S. Barry, D. Bayard, C. Beichman, M. Beltran, R. Bennett, P. Beyer, K. Bilby, D. Bliss, G. Bonfiglio, M. Bothwell, J. Bottenfield, D. Boussalis, C. Boyles, M. Brown, P. Brugarolas, R. Bunker, C. Cagle, C. Carrion, J. Casani, E. Cherniack, E. Clark, D. Cole, J. Craft, J. Cruz, S. Dekany, M. Deutsch, J. Dooley, R. Dumas, M. Ebersole, P. Eisenhardt, C. Elachi, W. Ellery, D. Elliott, K. Erickson, J. Evans, J. Fanson, T. Feehan, R. Fragoso, L. Francis, D. Gallagher, M. Gallagher, G. Ganapathi, M. Garcia, N. Gautier, T. Gavin, S. Giacoma, J. Gilbert, L. Gilliam, C. Glazer, P. Gluck, V. Gorjian, G. Greanias, C. Guernsey, A. Guerrero, M. Hashemi, G. Havens, C. Hidalgo, E. Higa, G. Hill, J. Hodder, H. Hotz, W. Hu, J. Hunt, Jr., D. Hurley, J. Ibanez, W. Irace, K. Jin, M.

---

1. This list is adapted from the Spitzer Handbook, 2010.

Johansen, M. Jones, B. Kahr, J. Kahr, B. Kang, P. Kaskiewicz, J. Keene, D. Kern, T. Kia, M. Kline, B. Korechoff, P. Kwan, J. Kwok, H. Kwong-Fu, M. Larson, M. Leeds, R. Lineaweaver, S. Linick, P. Lock, W. Lombard, S. Long, T. Luchik, J. Lumsden, M. Lysek, G. Macala, S. Macenka, N. Mainland, E. Martinez, M. McAuley, J. Mehta, P. Menon, R. Miller, C. Miyamoto, W. Moore, F. Morales, R. Morris, A. Nakata, B. Naron, A. Nash, D. Nichols, M. Osmolovsky, K. Owen-Mankovich, K. Patel, S. Peer, J. Platt, N. Portugues, D. Potts, S. Ramsey, S. Rangel, R. Reid, J. Reimer, E. Rice, D. Rockey, E. Romana, C. Rondeau, A. Sanders, M. Sarrel, V. Scarffe, T. Scharton, H. Schember, C. Scott, P. K. Sharma, T. Shaw, D. Shebel, J. Short, L. Simmons, C. Simon, B. Smith, R. Smith, P. Sorci, T. Specht, R. Spehalski, G. Squibb, S. Stanboli, K. Stapelfeldt, D. Stern, K. Stowers, J. Stultz, M. Tankenson, N. Thomas, R. Thomas, F. Tolivar, R. Torres, R. Tung, N. Vandermey, P. Varghese, M. Vogt, V. Voskanian, B. Waggoner, L. Wainio, T. Weise, J. Weiss, K. Weld, R. Wilson, M. Winters, S. Wissler, G. Yankura, K. Yetter.

Spitzer Science Center, California Institute of Technology (science operations): W. Amaya, L. Amy, P. Appleton, D. Ardila, L. Armus, J. Aronsson, D. Avila, M. Barba, S. Barba, J. Bauer, R. Beck, C. Bennett, J. Bennett, B. Bhattacharya, M. Bicay, C. Bluehawk, C. Boyles, H. Brandenburg, I. Bregman, C. Brinkworth, T. Brooke, J. Bruher, M. Burgdorf, S. Carey, M. Castillo, R. Chary, J. Chavez, W. Clavin, J. Colbert, S. Comeau, M. Crane, D. Daou, A. Dean, V. Desai, M. Dobard, S. Dodd, R. Ebert, R. Estrada, D. Fadda, S. Fajardo-Acosta, F. Fang, J. Fowler, D. Frayer, L. Garcia, C. Gelino, W. Glaccum, T. Goldina, W. Green, T. Greene, C. Grillmair, E. Ha, E. Hacopians, T. Handley, B. Hartley, I. Heinrichsen, G. Helou, S. Hemple, D. Henderson, L. Hermans, T. Hesselroth, A. Hoac, D. Hoard, R. Hoban, J. Howell, H. Hu, H. Hurt, H. Huynh, M. Im, J. Ingalls, E. Jackson, J. Jacobson, T. Jarrett, G. Johnson-McGee, J. Keller, A. Kelly, E. Kennedy, I. Khan, D. Kirkpatrick, S. Kolhatkar, J. Krick, M. Lacy, R. Laher, S. Laine, J. Lampley, W. Latter, T. Lau, W. Lee, M. Legassie, D. Levine, J. Li, P. J. Llamas, T. Lo, W. Lockhart, L. Ly, P. Lowrance, N. Lu, J. Ma, W. Mahoney, D. Makovoz, V. Mannings, F. Marleau, T. Marston, F. Masci, H. McCallon, B. McCollum, D. McElroy, M. McElveney, N. McElveney, V. Meadows, Y. Mei, S. Milanian, D. Mittman, A. Molloy, P. Morris, M. Moshir, R. Narron, B. Nelson, R. Newman, A. Noriega-Crespo, J. Ochotorena, P. Ogle, J. O'Linger, D. Padgett, R. Paladini, P. Patterson, A. Pearl, M. Pesenson, S. Potts, T. Pyle, W. Reach, L. Rebull, J. Rector, J. Rho, T. Roby, E. Ryan, R. Scholey, E. Scire, S. Shenoy, K. Sheth, A. Shields, D. Shupe, N. Silbermann, T. Soifer, I. Song, G. Squires, J. Stauffer, J. Stuesser, S. Stolovy, L. Storrie-Lombardi, J. Surace, H. Teplitz, M. Thaller, G. Turek, S. Tyler, S. Van Dyk, L. Vu, S. Wachter, C. Waterson, W. Wheaton, S. Wheelock, J. White, A. Wiercigroch, G. Wilson, X. Wu, L. Yan, F. Yu.

Construction

Ball Aerospace (Cryogenic Telescope Assembly): R. Abbott, D. Adams, S. Adams, J. Austin, B. Bailey, H. Bareiss, J. Barnwell, T. Beck, B. Benedict, M. Bilkey, W. Blalock, M. Breth, R. Brown, D. Brunner, D. Burg, W. Burmester, S. Burns, M. Cannon, W. Cash, T. Castetter, M. Cawley, W. Cebula, D. Chaney, G. J. Chodil, C. Cliff, S. Conley, A. Cooper, J. Cornwell, Sr., L. Cortelyou, J. Craner, K. Craven, D. Curtis, F. Davis, J. Davis, C. Dayton, M. Denaro, A. DiFronzo, T. Dilworth, N. Dobbins, C. Downey, A. Dreher, R. Drewlow, B. Dubrovin, J. Duncan, D. Durbin, S. Engles, P. Finley, J. Fleming, S. Forrest, R. Fredo, K. Gause, M. Gee, S. Ghesquiere, R. Gifford, J. Good, M. Hanna, D. Happs, F. Hausle, G. Helling, D. Herhager,

B. Heurich, E. Hicks, M. Hindman, R. Hopkins, H. Hoshiko, Jr., J. Houlton, J. Hueser, J. Hurt, W. Hyatt, K. Jackson, D. Johnson, G. Johnson, P. Johnson, T. Kelly, B. Kelsic, S. Kemper, R. Killmon, R. Knewtson, T. Konetski, B. Kramer, R. Kramer, L. Krauze, T. Laing, R. LaPointe, J. Lee, D. Lemon, P. Lien, R. Lytle, L. Madayev, M. Mann, R. Manning, J. Manriquez, M. Martella, G. Martinez, T. McClure, C. Meier, B. Messervy, K. Modafferi, S. Murray, J. Necas, Jr., M. Neitenbach, P. Neuroth, S. Nieczkoski, G. Niswender, E. Norman-Gravseth, R. Oonk, L. Oystol, J. Pace, K. Parrish, A. Pearl, Jr., R. Pederson, S. Phanekham, C. Priday, B. Queen, P. Quigley, S. Rearden, M. Reavis, M. Rice, M. Richardson, P. Robinson, C. Rowland, K. Russell, W. Schade, R. Schildgen, C. Schroeder, G. Schultz, R. Schweickart, J. Schweinsberg, J. Schwenker, S. Scott, W. Seelig, L. Seide, K. Shelley, T. Shelton, J. Shykula, J. Sietz, J. Simbai, L. Smeins, K. Sniff, B. Snyder, B. Spath, D. Sterling, N. Stoffer, B. Stone, M. Taylor, R. Taylor, D. Tennant, R. Tio, P. True II, A. Urbach, S. Vallejo, K. Van Leuven, L. Vernon, S. Volz, V. VonRuden, D. Waldeck, J. Wassmer, B. Welch, A. Wells, J. Wells, T. Westegard, C. Williamson, E. Worner, Jr., T. Yarnell, J. Yochum, A. Youmans, J. Zynsky.

Lockheed Martin (spacecraft, systems engineering, spacecraft operations): B. Adams, J. Akbarzadeh, K. Aline, T. Alt, G. Andersen, J. Arends, F. Arioli, A. Auyeung, D. Bell, R. Bell, F. Bennett, J. Bennett, M. Berning, H. Betts, M. Billian, S. Broadhead, B. Bocz, G. Bollendonk, N. Bossio, P. Boyle, T. Bridges, C. Brink, R. Brookner, J. Brunton, D. Bucher, M. Burrack, R. Caffrey, S. Carmer, P. Carney, T. Carpenter, R. Castro, J. Cattrysse, J. Cernac, G. Cesarone, K. Chan, C. Chang, M. Chuang, D. Chenette, A. Chopra, Z. Chou, W. Christensen, K. Chu, W. Clark, J. Clayton, S. Cleland, W. Clements, C. Colborn, A. Cooprider, B. Corwin, B. Costanzo, D. Cortes, M. Cox, M. Cox, J. Coyne, S. Curtin, G. Dankiewicz, C. Darr, J. Dates, J. Day, S. DeBrock, T. Decker, R. Defoe, J. Delavan, G. Delezynski, J. Delk, B. Dempsey, R. Dodder, T. Dougherty, H. Drosdat, G. Du, B. Dudginski, M. Dunn, R. Dunn, M. Dunnam, D. Durant, D. Eckart, B. Edwards, M. Effertz, L. Ellis, P. Emig, N. Etling, M. Etz, N. Fernando, C. Figge, R. Finch, S. Finnell, A. Fisher, M. Fisher, P. Fleming, D. Ford, K. Foster, J. Frakes, P. Frankel, D. Fulton, P. Galli, D. Garcia, M. Gardner, B. Garner, S. Gaskin, S. Gasner, M. Geil, E. Georgieva, T. Gibson, B. Goddard, M. Gonzalez, D. Goold, D. Graves, S. Gray, I. Grimm, J. Grinwis, M. Gronet, R. Grubic, S. Guyer, M. Haggard, J. Harrison, G. Hauser, C. Hayashi, P. Headley, W. Hegarty, S. Heires, J. Herrerias, D. Hirsch, K. Hooper, J. Horwath, S. Housten, D. Howell, L. Huff, G. Idemoto, B. Jackson, K. Janeiro, K. Johnson, M. Johnson, R. Kaiser, P. Kallemeyn, G. Kang, R. Kasuda, M. Kawasaki, B. Keeney, J. Kenworthy, C. King, A. Klavins, K. Klein, C. Klien, P. Klier, C. Koch, L. Koch, D. Koide, R. Kriegbaum, J. Kuchera, J. Ladewig, D. Lance, M. Lang, K. Lauffer, A. Lee, E. Lee, J. Lee, R. Lee, D. Leister, K. Loar, A. Lott, C. Love, N. Iyengar, P. Ma, A. Magallanes, A. Mainzer, T. Maloney, S. Mar, B. Marquardt, M. Martin, G. Mason, R. Maxwell, R. May, G. McAllister, S. McElheny, M. McGee, J. McGowan, D. McKinney, A. McMechen, E. Merlo, C. Mifsud, J. Miles, S. Miller, A. Minter, C. Miran, S. Mittal, R. Mock, R. Mock, J. Montgomery, J. Moore, H. Mora, M. Moradia, L. Morales, R. Morales, G. Morison, J. Mota, F. Moules, S. Mumaw, L. Naes, A. Nalbandian, J. Nelson, L. Nenoff, J. Neuman, D. Nguyen, K. Nguyen, T. Nguyen, D. Niebur, D. Nishimura, M. Ochs, T. Oliver, J. Oo, J. Ortiz, G. Pace, L. Padgett, N. Page, G. Painter, H. Pandya, L. Pappas, N. Pemberton, R. Peterson, H. Phan, L. Phan, J. Pine, R. Poling, R. Potash, D. Radtke, W. Ramos, T. Ransom, M. Ratajczyk, D. Read, S. Ready, M. Rich, R. Richey, H. Rizvi,

C. Rollin, C. Rudy, M. Rugebregt, R. Russek, B. Sable, C. Sandwick, M. Santos, N. Schieler, J. Schirle, G. Schlueter, M. Schmitzer, E. Sedivy, R. Seeders, S. Selover, R. Shaw, F. Sheetz, D. Shelton, R. Sherman, T. Sherrill, O. Short, R. Sison, B. Smith, F. Smith, S. Smith, B. Sotak, S. Spath, J. St. Pierre, K. Starnes, K. Stowers, J. Straetker, T. Stretch, S. Sulak, W. Sun, D. Swanson, C. Tatro, M. Tebo, D. Telford, A. Tessaro, J. Tietz, D. Tenerelli, J. Tolomeo, S. Toro-Allen, J. Tousseau, R. Traber, M. Tran, P. Travis, K. Uselman, S. Utke, N. Vadlamudi, R. VanBezooijen, J. Vantuno, R. Vasquez, G. Vergho, C. Voth, B. Vu, P. Wagner, M. White, M. Whitten, J. Wood, C. Worthley, D. Wright, C. Yanari, L. Yeaman, D. Zempel, S. Zeppa.

Infrared Array Camera (IRAC)

Smithsonian Astrophysical Observatory (project management and science): L. Allen, C. Arabadjis, M. Ashby, P. Barmby, V. Bawdekar, J. Boczenowski, D. Boyd, J. Campbell-Cameron, J. Chappell, M. Cohen, K. Daigle, L. Deutsch, L. Frazier, T. Gauron, J. Gomes, J. Hora, M. Horan, J. Huang, J. Huchra, E. Johnston, M. Kanouse, S. Kleiner, D. Koch, M. Marengo, S. Megeath, G. Melnick, W. Martell, P. Okun, M. Pahre, B. Patten, J. Polizotti, J. Rosenberg, H. Smith, J. Spitzak, R. Taylor, E. Tollestrup, J. Wamback, Z. Wang, S. Willner; NASA/ARC (Si:As detector array testing): J. Estrada, R. Johnston, C. McCreight, M. McKelvey, R. McMurray, R. McHugh, N. Moss, W. Oglivie, N. Scott, S. Zins.

NASA/GSFC (instrument development): T. Ackerson, M. Alexander, C. Allen, R. Arendt, M. Armbruster, S. Babu, W. Barber, R. Barney, L. Bashar, C. Bearer, C. Bernabe, W. Blanco, R. Boyle, K. Brenneman, G. Brown, M. Brown, G. Cammarata, S. Casey, P. Chen, M. Cushman, P. Davila, M. Davis, M. Dipirro, C. Doria-Warner, W. Eichhorn, D. Evans, D. Fixsen, J. Florez, J. Geiger, D. Gezari, D. Glenar, J. Golden, P. Gorog, S. Graham, C. Hakun, P. Haney, T. Hegerty, M. Jhabvala, F. Jones-Selden, R. Jungo, G. Karpati, R. Katz, R. Kichak, R. Koehler, R. Kolecki, D. Krebs, A. Kutyrev, J. Lander, M. Lander, N. Lee, J. Lohr, P. Losch, J. MacLoed, R. Maichle, S. Mann, N. Martin, P. Maymon, D. McComas, J. McDonnell, D. McHugh, J. Mills, C. Moiser, S. Moseley, T. Nguyen, T. Powers, K. Rehm, G. Reinhardt, J. Rivera, F. D. Robinson, C. Romano, M. Ryschkewitsch, S. Schwinger, K. Shakoorzadeh, P. Shu, N. Shukla, S. Smith, R. Stavely, W. Tallant, V. Torres, C. Trout, C. Trujillo, D. Vavra, G. Voellmer, V. Weyers, R. Whitley, J. Wolfgang, L. Workman, D. Yoder.

Raytheon Vision Systems (detector arrays): C. Anderson, J. Asbrock, V. Bowman, G. Chapman, E. Corrales, G. Domingo, A. Estrada, B. Fletcher, A. Hoffman, L. Lum, N. Lum, S. Morales, O. Moreno, H. Mosel-Riedo, J. Rosbeck, K. Schartz, M. Smith, S. Solomon, K. Sparkman, P. Villa, S. Woolaway.

University of Arizona: W. Hoffman, T. Tysenn, P. Woida.

University of Rochester (InSb detector array testing): C. Bacon, R. Benson, H. Chen, J. Comparetta, N. Cowen, M. Drennan, W. Forrest, J. Garnett, B. Goss, S. Libonate, R. Madson, B. Marazus, K. McFadden, C. McMurtry, D. Myers, Z. Ninkov, R. Overbeck, J. Pipher, R. Sarkis, J. Schoenwald, B. White, J. Wu.

Infrared Spectrograph (IRS)

Cornell University (project management and science): D. Barry, S. V. W. Beckwith, J. Bernhard-Salas, C. Blacken, B. Brandl, V. Charmandaris, M. Colonno, S. Corbin, P. Devine, D. Devost, J. Diller, K. Duclos, E. Furlan, G. Gull, P. Hall, L. Hao, C. Henderson, T. Herter, J. Higdon,

S. Higdon, P. Howell, L. McCall, A. Parks, B. Pirger, A. Rakowski, S. Reinehart, A. Reza, E. E. Salpeter, J. Schoenwald, G. Sloan, J. Smith, H. Spoon, K. Uchida, D. Weedman, J. Wilson.

University of Rochester: D. M. Watson, W. F. Forrest.

California Institute of Technology: K. Matthews.

Ball Aerospace (instrument development): D. Alderman, D. Anthony, M. Bangert, J. Barnwell, A. Bartels, S. Becker, W. Belcher, J. Bergstrom, D. Bickel, M. Bolton, S. Burcar, D. Burg, S. Burns, S. Burns, D. Burr, P. Burrowes, W. Cebula, C. Conger, J. Crispin, M. Dean, M. D'Ordine, S. Downey, R. Drewlow, L. Duchow, D. Eva, C. Evans, M. Foster, S. Fujita, D. Gallagher, A. Gaspers, P. Gentry, S. Giddens, J. Graw, M. Hanna, A. Haralson, M. Henderson, D. Herhager, J. Hill, S. Horacek, M. Huisjen, S. Hunter, J. Jacob, R. Karre, L. Larsen, P. Lien, R. Manning, J. Marriott, D. McConnell, M. McIntosh, R. McIntosh, G. Mead, B. Michelson, B. Miller, J. Moorehead, M. Morris, J. Murphy, M. Nelson, J. Pacha, I. Patrick, A. Pearl, B. Pett, S. Randall, C. Rowland, R. Sandoval, D. Sealman, K. Shelley, J. Simbai, L. Smeins, C. Stewart, G. Taudien, D. Tennant, J. Troeltzsch, B. Unruh, J. van Cleve, C. Varner, J. Winghart, J. Workman; Rockwell (detector arrays): B. Beardwood, J. Huffman, D. Reynolds, D. Seib, M. Stapelbroek, S. Stetson.

OCLI (filters): S. Corda, B. Dungan, D. Favot, S. Highland, M. Inong, V. Jauregui, C. Kennemore, B. Langley, S. Mansour, R. Mapes, M. Mazzuchi, C. Piazzo.

Multiband Imaging Photometer for Spitzer (MIPS)

University of Arizona (project management, array construction, and science): A. Alonso-Herrero, M. Alwardi, I. Barg, M. Blaylock, M. Bradley, M. Buglewicz, J. Cadien, A. Churchill, H. Dang, L. Davidson, J. T. Davis, H. Dole, E. Egami, C. Engelbracht, K. A. Ennico, J. Facio, J. Flores, K. D. Gordon, L. Hammond, D. Hines, J. Hinz, R. Hodge, T. Horne, P. Hubbard, D. M. Kelly, D. Knight, K. A. Kormos, E. LeFloc'h, F. J. Low, M. McCormick, T. J. McMahon, T. Milner, K. Misselt, J. Morrison, K. Morse, J. Muzerolle, G. X. Neugebauer, L. Norvelle, C. Papovich, P. Perez-Gonzalez, M. J. Rieke, G. Rivlis, P. Rogers, R. Schnurr, M. Scutero, C. Siqueiros, P. Smith, J. A. Stansberry, P. Strittmatter, K. Su, C. Thompson, P. van Buren, S. Warner, K. White, D. A. Wilson, G. S. Winters, E. Young.

University of California, Berkeley/LBNL (science and detector material): E. Arens, J. W. Beeman, E. E. Haller, P. L. Richards.

Jet Propulsion Laboratory (science): C. Beichman, K. Stapelfeldt.

National Optical Astronomy Observatories (science): J. Mould.

Center for Astrophysics (science): C. Lada; Ball Aerospace (instrument development): D. Bean, M. Belton, T. Bunting, W. Burmester, S. Castro, C. Conger, L. Derouin, C. Downey, B. Frank, H. Garner, P. Gentry, T. Glenn, M. Hegge, G. B. Heim, M. L. Henderson, F. Lawson, K. MacFeely, B. McGilvray, R. Manning, D. Michika, C. D. Miller, D. Morgan, M. Neitenbach, R. Novaria, R. Ordonez, R. J. Pearson, Bruce Pett, K. Rogers, J. P. Schwenker, K. Shelley, S. Siewert, D. W. Strecker, S. Tennant, J. Troeltzsch, B. Unruh,

R. M. Warden, J. Wedlake, N. Werholz, J. Winghart, R. Woodruff, C. Yanoski.

Raytheon (readout development): J. Asbrock, A. Hoffmann, N. Lum.

Ames Research Center (readout development): C. McCreight.

QM Industries (far-infrared filters): P. A. R. Ade.

Blackforest Engineering (engineering support): S. Gaalema.

Battel Engineering (engineering support): S. Battel.

SRON (scan mirror development): T. Degraauw.

Bibliography

This is a list of all the source materials referenced in preparing this monograph. Primary source materials (such as interviews and meeting minutes collected by the author) will be available from the Spitzer Project Monograph archive at NASA's History Division. Secondary source materials (such as books, journal articles, and government reports) are generally available in libraries or online (links are provided where available).

Government Reports

Advisory Committee on the Future of the U.S. Space Program (Norman Augustine, chair), *Report of the Advisory Committee on the Future of the U.S. Space Program*, NTRS 19910012869 (Washington, DC: NASA, 1990), a.k.a. the "Augustine report."

Air Force Geophysics Laboratory (AFGL) Sky Survey team (Stephan D. Price and Russell G. Walker, lead authors), *The AFGL Four Color Infrared Sky Survey: Catalog of Observations at 4.2, 11.0, 19.8, and 27.4 microns*, AFGL-TR-76-0208 (Hanscom AFB, MA: AFGL, 1976).

Ames Research Center, *Lear-Jet Airborne Observatory Investigators Handbook*, NASA-TM-108623 (Moffett Field, CA: NASA Ames, 1974).

Ames Research Center (Craig R. McCreight, ed.), *Proceedings of the 3rd Infrared Detector Technology Workshop*, NASA-TM-102209, NTRS 19900011997 (Moffett Field, CA: NASA Ames, 1989).

Astronomy Survey Committee (Jesse Greenstein, chair), *Astronomy and Astrophysics for the 1970's, Volume 1: Report of the Astronomy Survey Committee and Volume 2: Reports of the Panels* (Washington, DC: National Academy of Sciences, 1972), a.k.a. the "Greenstein report"; vol. 1 available at *http://www.nap.edu/catalog. php?record_id=13231* and vol. 2 at *http://www. nap.edu/catalog/21812/astronomy-and-astrophysics-for-the-1970s-volume-2-reports-of* (accessed 30 August 2016).

Astronomy Survey Committee (George B. Field, chair), *Astronomy and Astrophysics for the 1980's, Volume 1: Report of the Astronomy Survey Committee* and *Volume 2: Reports of the Panels* (Washington, DC: National Academy of Sciences, 1982), a.k.a. the "Field report"; vol. 1 available at *http://www.nap.edu/catalog.php?record_id=549* and vol. 2 at *http://www.nap.edu/catalog.php?record_id=550* (accessed 30 August 2016).

Astronomy and Astrophysics Survey Committee (John Bahcall, chair), *The Decade of Discovery in Astronomy and Astrophysics (*Washington, DC: National Academy Press, 1991), a.k.a. the "Bahcall report"; available at *http://www.nap.edu/catalog.php?record_id=1634* (accessed 30 August 2016).

Columbia Accident Investigation Board (Hal W. Gehman, Jr., chair), *The Columbia Accident Investigation Board Report*, 6 vols. (Washington, DC: Government Printing Office, 2003), available at *http://www.nasa.gov/columbia/caib/html/start.html* (accessed 30 August 2016).

Congressional Research Service (David P. Radzanowski), *NASA Under Scrutiny: The Shuttle and Hubble*, Issue Brief no. IB90114, 13 September 1990 (Washington, DC: Congressional Research Service, 1990).

Congressional Research Service (Marcia S. Smith), *Space Launch Vehicles: Government Activities, Commercial Competition, and Satellite Exports*, Issue Brief no. IB93062, 3 February 2003 (Washington, DC: Congressional Research Service, 2003).

Congressional Research Service (Bill Heniff, Jr., Megan Lynch, and Jessica Tollestrup), *Introduction to the Federal Budget Process*. Report no. 98-721, 3 December 2012 (Washington, DC: Congressional Research Service, 2012).

FIRST: Focal Plane Instruments and Requirements Science Team (Fred C. Witteborn and David M. Rank, cochairs), *Appendices to FIRST Interim Report on SIRTF*, 14 April 1978 (Moffett Field, CA: Ames Research Center, 1978); to be deposited in NASA HRC.

Goddard Space Flight Center, *Final Report of the Space Shuttle Payload Planning Working Groups, Volume 1: Astronomy*, NASA-TM-X-69409 (Greenbelt, MD: NASA, 1973).

Goddard Space Flight Center, *Orbiting Solar Observatory: OSO I: The Project Summary*, SP-57 (Greenbelt, MD: NASA, 1965).

Goddard Space Flight Center, *Proceedings of the Space Shuttle Sortie Workshop, Volume 1: Policy and System Characteristics* and *Volume 2: Working Group Reports*, NASA-TM-X-68841 (vol. 1) and NASA-TM-X-68842 (vol. 2), (Greenbelt, MD: NASA, 1972).

Great Observatories for Space Astrophysics, The, 1st edition, CR-176754, NTRS 19860015241 (Washington, DC: NASA, 1985).

Great Observatories for Space Astrophysics, The, 2nd edition, NP-128, NTRS 19920001848 (Washington, DC: NASA, 1991).

Mars Observer Mission Failure Investigation Board (Timothy P. Coffey, chair), *Mars Observer Mission Failure Investigation Board Report* (Washington, DC: NASA, 1993), available at *http://spacese.spacegrant.org/Failure%20Reports/Mars_Observer_12_93_MIB.pdf* (accessed 30 August 2016).

Martin Marietta study group (L. E. Bareiss, R. O. Rantanen, and E. B. Ress), *Payload/Orbiter Contamination Control Requirement Study Final Report*, MRC 74-93 (Denver, CO: Martin Marietta Aerospace Corp., 1974).

NASA Office of the Inspector General, *Hubble Space Telescope Cost-Saving Initiatives*, Audit Report IG-99-013, available at *http://www.hq.nasa.gov/office/hqlibrary/documents/o43731805.pdf* (accessed 30 August 2016).

Neugebauer, Gerry, and Robert D. Leighton, *Two-Micron Sky Survey: A Preliminary Catalog*, SP-3047, NTRS 19690028611 (Washington, DC: NASA, 1969).

Space Science Board (SSB) Study Group, *A Review of Space Research: The Report of the Summer Study Conducted Under the Auspices of the Space Science Board of the National Academy of Sciences*, NRCP 1079 (Washington, DC: National Research Council, 1962), a.k.a. the "SSB Study Group, 1962" or the "Space Science Summer Study."

Space Science Board (SSB) Study Group, *Space Research: Directions for the Future; Report of a Study by the Space Science Board, Woods Hole, Massachusetts, 1965*, NRCP 1403 (Washington, DC: National Research Council, 1966), a.k.a. the "SSB Study Group, 1966."

Space Science Board (SSB) Study Group (John Findlay, chair), *Scientific Uses of the Space Shuttle*, NTRS 19740019254 (Washington, DC: National Academy of Sciences, 1974).

Space Science Board (office of the Chairman), *Policy Positions on (1) Man's Role in the National Space Program and (2) Support of Basic Research for Space Science*, letter from Lloyd V. Berkner to James E. Webb, NASA Administrator, 31 March 1961, available at *http://www.nap.edu/catalog/12427/policy-positions-on-1-mans-role-in-the-national-space-program-and-2-support-of-basic-research-for-space-science-march-27* (accessed 30 August 2016).

Space Shuttle Payloads: Hearing Before the U.S. Senate Committee on Aeronautical and Space Sciences, 30–31 October 1973, 93rd Congress, 1st Session (Washington, DC: U.S. Government Printing Office, 1973).

Historical Monographs and Edited Collections

Bugos, Glenn E., *Atmosphere of Freedom: Sixty Years at the NASA Ames Research Center*, SP-4314 (Washington, DC: NASA, 2000), available at *http://history.nasa.gov/SP-4314/sp4314.htm* (accessed 30 August 2016).

Cruikshank, Dale P., "Gerard Peter Kuiper," *Biographical Memoirs*, vol. 62 (Washington, DC: National Academies Press, 1993), pp. 266–273.

De Vaucouleurs, Gérard H., "Harold Lester Johnson," *Biographical Memoirs*, vol. 67 (Washington, DC: National Academies Press, 1995), pp. 243–261.

Gillespie, Carl, ed., *Gerard P. Kuiper Airborne Observatory and Learjet Observatory Plus Astronomy-Related Publications from the NASA Convair 990 Aircraft* (Moffett Field, CA: Ames Research Center, 1991).

Launius, Roger D., and Aaron K. Gillette, *Toward a History of the Space Shuttle: An Annotated Bibliography* (Washington, DC: NASA Monographs in Aerospace History, 1992), available at *https://www.nasa.gov/sites/default/files/708235main_Shuttle_Bibliography_1-ebook.pdf* (accessed 30 August 2016).

Logsdon, John M., ed. (with Amy Paige Snyder, Roger D. Launius, Stephen J. Garber, and Regan Anne Newport), *Exploring the Unknown: Selected Documents in the History of the U.S. Civil Space Program*, vol. 5, *Exploring the Cosmos*, SP-2001-4407 (Washington, DC: NASA, 2001).

Neal, Valerie, Tracy McMahan, and Dave Dooling, eds., *Science in Orbit: The Shuttle & Spacelab Experience, 1981–1986*, NP-119 (Huntsville, AL: NASA, 1988), available at *http://history.nasa.gov/NP-119/contents.htm* (accessed 30 August 2016).

Newell, Homer E., *Beyond the Atmosphere: Early Years of Space Science*, SP-4211 (Washington, DC: NASA, 1980), chap. 17; available at *http://history.nasa.gov/SP-4211/cover.htm* (accessed 30 August 2016).

Rumerman, Judy A., ed., *NASA Launch Systems, Space Transportation, Human Spaceflight, and Space Science, 1979–1988*, vol. 5 of *NASA Historical Data Books*, SP-4012 (Washington, DC: NASA, 1999). The chapter on space science is available at *http://history.nasa.gov/SP-4012/vol5/vol_v_ch_4.pdf* (accessed 30 August 2016).

Rumerman, Judy A., ed., *NASA Launch Systems, Space Transportation, Human Spaceflight, and Space Science, 1989–1998*, vol. 7 of *NASA Historical Data Books*, SP-4012 (Washington, DC: NASA, 2009). The chapter on space science is available at *http://history.nasa.gov/SP-4012v7ch4.pdf* (accessed 30 August 2016).

Books

Allen, Thomas, *Managing the Flow of Innovation* (Cambridge, MA: MIT Press, 1997).

Baumgartner, Frank R., and Bryan D. Jones, *Agendas and Instability in American Politics*, 2nd ed. (Chicago: IL: University of Chicago Press, 2009).

Bijker, Wiebe E., Thomas P. Hughes, and Trevor J. Pinch (eds.), *The Social Construction of Technological Systems: New Directions in the Sociology and History of Technology* (Cambridge, MA: MIT Press, 1987).

Duerbeck, Hilmar W., "German Astronomy in the Third Reich," in *Organizations and Strategies in Astronomy*, vol. 7, ed. André Heck (Dordrecht, Netherlands: Springer, 2006).

Florence, Ronald, *The Perfect Machine: Building the Palomar Telescope* (New York: Harper Perennial, 1995).

Friedman, Norman, *The Naval Institute Guide to World Naval Weapons Systems*, 5th ed. (Annapolis, MD: U.S. Naval Institute Press, 2006).

Harwit, Martin O., *Cosmic Discovery: The Search, Scope and Heritage of Astronomy* (New York, NY: Basic Books, 1981).

Harwit, Martin O., "The Early Days of Infrared Space Astronomy" in *The Century of Space*

Science, vol. 1, ed. Johan A. M. Bleeker et al. (Dordrecht, Netherlands: Kluwer, 2001).

Jones, Bryan D., and Frank R. Baumgartner, *The Politics of Attention: How Government Prioritizes Problems* (Chicago: University of Chicago Press, 2005).

Kingdon, John W., *Agendas, Alternatives, and Public Policies*, 2nd ed. (New York: Longman, 1995).

Mather, John C., and John Boslough, *The Very First Light: The True Inside Story of the Scientific Journey Back to the Dawn of the Universe*, rev. ed. (New York: Basic Books, 2008).

McCurdy, Howard E., *Faster, Better, Cheaper: Low-Cost Innovation in the U.S. Space Program* (Baltimore, MD: Johns Hopkins University Press, 2001).

McDougall, Walter A., *...the Heavens and the Earth: A Political History of the Space Age* (Baltimore, MD: Johns Hopkins Univ. Press, 1985), especially chapter 19.

Neufeld, Michael J., *The Rocket and the Reich: Peenemünde and the Coming of the Ballistic Missile Era* (Cambridge, MA: Harvard University Press, 1996).

Pellerin, Charles J., *How NASA Builds Teams: Mission Critical Soft Skills for Scientists, Engineers, and Project Teams* (Hoboken, NJ: Wiley, 2009).

Rieke, George H., *The Last of the Great Observatories: Spitzer and the Era of Faster, Better, Cheaper at NASA* (Tucson: University of Arizona Press, 2006).

Singh, Simon, *Big Bang: The Origin of the Universe* (New York: Harper Perennial, 2005).

Smith, Robert W., with contributions by Paul A. Hanle, Robert Kargon, and Joseph N. Tatarewicz, *The Space Telescope: A Study of NASA, Science, Technology, and Politics* (Cambridge, U.K.: Cambridge University Press, 1993).

Tatarewicz, Joseph N., *Space Technology and Planetary Astronomy* (Bloomington, IN: Indiana University Press, 1990).

Tucker, Wallace, and Karen Tucker, *Revealing the Universe: The Making of the Chandra X-Ray Observatory* (Cambridge, MA: Harvard University Press, 2001).

Van Allen, James A., *Origins of Magnetospheric Physics* (Iowa City, IA: University of Iowa Press, 2004).

Journal Articles and Conference Proceedings

Audia, Pino, and Chris Rider, "A Garage and an Idea: What More Does an Entrepreneur Need?," *California Management Review* 48, no. 1 (2005): 6–28.

Becklin, Eric E., and Gerry Neugebauer, "Observations of an Infrared Star in the Orion Nebula," *Astrophysical Journal* 147 (1967): 799–802.

Bicay, Michael D., and Michael W. Werner, "SIRTF: Linking the Great Observatories with the Origins Program," *ASP Conference Series (ORIGINS)*, eds. Charles E. Woodward, J. Michael Shull, and Harley A. Thronson, Jr., vol. 148 (1998): 290–297.

Brandshaft, Donald, Robert A. McLaren, and Michael W. Werner, "Spectroscopy of the Orion Nebula from 80 to 135 microns," *Astrophysical Journal* 199, pt. 2 (1975): L115–L117.

Danielson, Robert E., John E. Gaustad, Martin Schwarzschild, Harold F. Weaver, and Neville J. Woolf, "Mars Observations from Stratoscope II," *Astronomical Journal* 69, no. 5 (1964): 344–352.

Davidson, Kris, and Martin O. Harwit, "Infrared and Radio Appearance of Cocoon Stars," *Astrophysical Journal* 148 (1967): 443–448.

Dicke, Robert H., P. James E. Peebles, Peter G. Roll, and David T. Wilkinson, "Cosmic Black-Body Radiation," *Astrophysical Journal* 142 (1965): 414–419.

Dolci, Wendy Whiting, "Milestones in Airborne Astronomy: From the 1920's to the Present," *AIAA, 1997 World Aviation Congress, October 13–16, 1997,* Anaheim, CA; available as NASA ARC publication 975609 and at *https://www.sofia.usra.edu/sites/default/files/97-Whiting_AeroHistory.pdf* (accessed 30 August 2016).

Fanson, James, Giovanni Fazio, James Houck, Tim Kelly, George Rieke, Domenick Tenerelli, and

Milt Whitten, "The Space Infrared Telescope Facility (SIRTF)," *Proceedings of the SPIE (Space Telescopes and Instruments)*, eds. Pierre Y. Bely and James B. Breckinridge, vol. 3356 (1998): 478–491.

Fazio, Giovanni G., "Planned NASA Space Infrared Astronomy Experiments," *Advances in Space Research* 2, no. 4 (1982): 97–106.

Fazio, Giovanni G., and Everett M. Hafner, "Directional High Energy Gamma-Ray Counter," *Review of Scientific Instruments* 32, no. 6 (1961): 697–702.

Fazio, Giovanni G., and Everett M. Hafner, "OSO-1 High-Energy Gamma-Ray Experiment," *Journal of Geophysical Research* 72, no. 9 (1967): 2452–2455.

Fazio, Giovanni G., "Flying High-Altitude Balloon-Borne Telescopes 50 Years Ago," *37th COSPAR Scientific Assembly*, held 13–20 July 2008 in Montréal, Canada (2009), p. 862.

Gehrz, Robert D., Thomas L. Roellig, Michael W. Werner, Giovanni G. Fazio, James R. Houck, Frank J. Low, George H. Rieke, B. Thomas Soifer, Deborah A. Levine, and Edward A. Romana, "The NASA Spitzer Space Telescope," *Review of Scientific Instruments* 78, article no. 011302 (2007): 1–38.

Giavalisco, Mauro, Henry C. Ferguson, Anton M. Koekemoer, Mark Dickinson, David M. Alexander, Franz E. Bauer, Jacqueline Bergeron, Carl Biagetti, W. Niel Brandt, Stefano Casertano, Catherine Cesarsky, Eleni Chatzichristou, Chris Conselice, Stefano Cristiani, L. Nicolaci Da Costa, Tomas Dahlen, Duilia de Mello, Peter Eisenhardt, Thomas Erben, S. Michael Fall, Chris Fassnacht, Robert Fosbury, Andrew Fruchter, Jonathan P. Gardner, Norman Grogin, Richard N. Hook, Ann E. Hornschemeier, Rafal Idzi, Shardha Jogee, Claudia Kretchmer, Vicki Laidler, Kyoung-Soo Lee, Mario Livio, Ray Lucas, Piero Madau, Bahram Mobasher, Leonidas A. Moustakas, Mario Nonino, Paolo Padovani, Casey Papovich, Yousin Park, Swara Ravindranath, Alvio Renzini, Marin Richardson, Adam Riess, Piero Rosati, Mischa Schirmer, Ethan Schreier, Rachel S. Somerville, Hyron Spinrad, Daniel Stern, Massimo Stiavelli, Lou Strolger, Claudia Megan Urry, Benoit Vandame, Robert Williams, and Chris Wolf, "The Great Observatories Origins Deep Survey (GOODS): Initial Results from Optical and Near-Infrared Imaging," *Astrophysical Journal* 600 (2004): L93–L98.

Hargadon, Andrew, and Robert I. Sutton, "Technology Brokering and Innovation in a Product Development Firm," *Administrative Science Quarterly* 42, no. 4 (1997): 716–749.

Harwit, Martin O., "The Number of Class A Phenomena Characterizing the Universe," *Royal Astronomical Society Quarterly Journal* 16, no. 4 (December 1975): 378–409.

Harwit, Martin O., "Conceiving and Marketing NASA's Great Observatories," *Experimental Astronomy* 26, nos. 1–3 (2009): 163–177.

Hayakawa, Satio, and Mamoru Saito, "Astronomy in Japan," *Astrophysics and Space Science* 99, nos. 1–2 (1984): 393–402.

Houck, James R., Thomas L. Roellig, Jeffrey Van Cleve, Bernhard R. Brandl, and Kevin I. Uchida, "IRS: The Spectrograph on SIRTF; Its Fabrication and Testing," *Proceedings of the SPIE (Infrared Spaceborne Remote Sensing VIII)*, ed. M. Strojnik and B. F. Andresen, vol. 4131 (2000): 70–77.

Jansky, Karl G., "Radio Waves from Outside the Solar System," *Nature* 132, no. 3323 (1933): 66.

Johnson, Harold L., "Infrared Stellar Photometry," *Astrophysical Journal* 135 (1962): 69–77.

Johnson, Harold L., "Astronomical Measurements in the Infrared," *Annual Review of Astronomy and Astrophysics* 4 (1966): 193–206.

Johnson, Harold L., and Frank J. Low, "Stellar Photometry at 10 µ," *Astrophysical Journal* 139 (1964): 1130–1134.

Kleinmann, Douglas E., and Frank J. Low, "Discovery of an Infrared Nebula in Orion," *Astrophysical Journal* 149 (1967): L1.

Kniffen, Donald A. "The Gamma-Ray Observatory," *Annals of the New York Academy of Sciences* 571 (1989): 482–496.

Kuiper, Gerard P., Wallace R. Wilson, and Robert J. Cashman, "An Infrared Stellar Spectrometer," *Astrophysical Journal* 106, no. 2 (1947): 243–250.

Lebreton, Jean-Pierre, and Dennis L. Matson, "The Huygens Probe: Science, Payload and Mission Overview," *Space Science Reviews* 104, no. 1 (2002): 59–100.

Lester, Dan, Jason Budinoff, and Charles Lillie, "Large Infrared Telescopes in the Exploration Era: SAFIR," *Proceedings of the SPIE (UV/Optical/IR Space Telescopes: Innovative Technologies and Concepts III)*, eds. Howard A. MacEwen and James B. Breckinridge, vol. 6687 (2007): 66870M–66870M.

Lock, Patricia, "SIRTF: Inheritance, Adaptation, and Advancement," *Proceedings of AIAA SpaceOps Conference* (June 2000), available at *http://hdl.handle.net/2014/14468*.

Low, Frank J., George H. Rieke, and Robert D. Gehrz, "The Beginning of Modern Infrared Astronomy," *Annual Review of Astronomy and Astrophysics* 45 (2007): 43–75.

Low, Frank J., "Low-Temperature Germanium Bolometer," *Journal of the Optical Society of America* 51, no. 11 (1961): 1300–1304.

McCarthy, Steve G., Lou S. Young, and Fred C. Witteborn, "A Large Cooled Infrared Telescope Facility for Spacelab," *Proceedings of the 21st Annual Meeting of American Astronautical Society (Space Shuttle Missions of the '80s)*, Denver, CO, 24–26 August 1975, NTRS 19760029854.

Meredith, Leslie H., Melvin B. Gottlieb, and James A. Van Allen, "Direct Detection of Soft Radiation Above 50 Kilometers in the Auroral Zone," *Physical Review* 97, no. 1 (1955): 201–205.

Miles, John W., Susan H. Linick, Stacia Long, John Gilbert, Mark Garcia, Carole Boyles, Michael Werner, and Robert K. Wilson, "Execution of the Spitzer In-Orbit Checkout and Science Verification Plan," *Proceedings of the SPIE (Astronomical Telescopes and Instrumentation)*, ed. John C. Mather, vol. 5475 (2004): 134–145.

Moore, W. J., and Henry Shenker, "A High-Detectivity Gallium-Doped Germanium Detector for the 40–120μ Region," *Infrared Physics* 5, no. 3 (1965): 99–106.

Morrison, Philip, "On Gamma-Ray Astronomy," *Il Nuovo Cimento* 7, no. 6 (1958): 858–865.

Münch, Guido, Gerry Neugebauer, and Dan McCammon, "Infrared Coronal Lines II: Observation of [SI x] $\lambda 1.43$ μ and [Mg VIII] $\lambda 3.03$ μ," *Astrophysical Journal* 149 (1967): 681–686.

Murray, Bruce C., and Robert L. Wildey, "Stellar and Planetary Observations at 10 Microns," *Astrophysical Journal* 137 (1963): 692–693.

Murray, Bruce C., Robert L. Wildey, and James A. Westphal, "Venus, a Map of Its Brightness Temperature," *Science* 140, no. 3565 (1963a): 391–392.

Murray, Bruce C., Robert L. Wildey, and James A. Westphal, "Infrared Photometric Mapping of Venus Through the 8–14 Micron Atmospheric Window," *Journal of Geophysical Research* 68 (1963b): 4813–4818.

Naugle, John E., "Research with the Space Shuttle," *Physics Today* 26, no. 11 (1973): 30–35, 37.

Neel, Jr., Carr B., Roy N. Griffin, and John P. Millard, "Studies Related to Satellite Thermal Control: Measurements of Earth-Reflected Sunlight and Stability of Thermal-Control Coatings," *Solar Physics* 6, no. 2 (1963): 235–240.

Neugebauer, Gerry, Dowell E. Martz, and Robert B. Leighton, "Observations of Extremely Cool Stars," *Astrophysical Journal* 142 (1965): 399–401.

Penzias, Arno A., and Robert W. Wilson, "A Measurement of Excess Antenna Temperature at 4080 Mc/s," *Astrophysical Journal* 142 (1965): 419–421.

Price, Stephan D., "Infrared Sky Surveys," *Space Science Review* 142, nos. 1–4 (2009): 233–321.

Reber, Grote, "Cosmic Static," *Astrophysical Journal* 100 (1944): 279–287.

Reichhardt, Tony, "The First Photo from Space," *Air & Space Magazine,* Smithsonian Institution (1 November 2006), available at *http://www.airspacemag.com/space-exploration/FEATURE-FirstPhoto.html* (accessed 30 August 2016).

Richards, Paul L., and Craig R. McCreight, "Infrared Detectors for Astrophysics," *Physics Today* 58, no. 2 (2005): 41–47.

Rogers, John H., "Origins of the Ancient Constellations: I. The Mesopotamian Traditions," *Journal of the British Astronomical Association* 108, no. 1 (1998): 9–28.

Roman, Nancy G., "A Note on Beta-Cephei," *Astrophysical Journal* 106, no. 2 (1947): 311–312.

Savedoff, Malcolm P., "The Crab and Cygnus A as Gamma Ray Sources," *Il Nuovo Cimento* 13, no. 1 (1959): 12–18.

Schwarzchild, Martin, and Barbara Schwarzchild, "Balloon Astronomy," *Scientific American* 200, no. 5 (1959): 52–59.

Schwenker, John P., Bernhard R. Brandl, William F. Hoffmann, Joseph L. Hora, Amanda K. Mainzer, John E. Mentzell, and Jeffrey Van Cleve, "SIRTF-CTA Optical Performance Test," *Proceedings of the SPIE (IR Space Telescopes and Instruments)*, ed. John C. Mather, vol. 4850 (2003): 304–317.

Sood, Ravinder K., "Detection of High-Energy Gamma-Rays from the Galactic Disk at Balloon Altitudes," *Nature* 222, no. 5194 (1969): 650–652.

Thronson, Harley A., John K. Davies, John Hackwell, Timothy G. Hawarden, Roger F. Knacke, Daniel Lester, C. Mattias Mountain, "EDISON: The Next Generation Infrared Space Observatory," *Space Science Reviews* 61, no. 1 (1992): 145–169.

Thronson, Harley A., Timothy G. Hawarden, Alan J. Penny, Laurent Vigroux, Gennadii Sholomitskii, "The Edison Infrared Space Observatory," *Space Science Reviews* 74, nos. 1–2 (1995): 139–144.

Tousey, Richard, "Solar Research from Rockets," *Science* 134, no. 3477 (1961): 441–448.

Townes, Charles H., "A Physicist Courts Astronomy," *Annual Review of Astronomy and Astrophysics* 35 (1997): xiii–xliv.

Van Allen, James A., "Direct Detection of Auroral Radiation with Rocket Equipment," *Proceedings of the National Academy of Sciences of the United States of America* 43, no. 1 (1957): 57–62.

Van Allen, James A., George H. Ludwig, Ernest C. Ray, and Carl. E. McIlwain, "Observations of High-Intensity Radiation by Satellites 1958 Alpha and Gamma," *Jet Propulsion* 28, no. 9 (1958): 588–592.

Wax, Robert L., "Balloon Observations of Infrared and X-Ray Intensities in the Auroral Zone," *Journal of Atmospheric and Terrestrial Physics* 28, no. 4 (1965): 397–407.

Werner, Michael W., Giovanni G. Fazio, George H. Rieke, Thomas L. Roellig, and Dan M. Watson, "First Fruits of the Spitzer Space Telescope: Galactic and Solar System Studies," *Annual Review of Astronomy and Astrophysics* 44 (2006): 269–321.

Werner, Michael W., "Spitzer: The First 30 Months," *Astronomy & Geophysics* 47, no. 6 (2006): 6.11–6.16.

Werner, Michael W., "A Short and Personal History of the Spitzer Space Telescope," *ASP Conference Series (The Spitzer Space Telescope: New Views of the Cosmos)*, eds. Lee Armus and William T. Reach, vol. 357 (2006), pp. 7–22; preprint available at *http://arxiv.org/PS_cache/astro-ph/pdf/0503/0503624v1.pdf* (accessed 30 August 2016).

Werner, Michael W., "Frank Low's Contributions to the History of Infrared Astronomy," presented at the 214th Meeting of the American Astronomical Society, 7–11 June 2009, Pasadena, CA.

Westphal, James A., Bruce C. Murray, and Dowell E. Martz, "An 8–14 Micron Infrared Astronomical Photometer," *Applied Optics* 2, no. 7 (1963): 749–775.

Wilson, Robert K., and Charles P. Scott, "The Road to Launch and Operations of the Spitzer Space Telescope," paper presented at the *SpaceOps Conference*, Rome, Italy, 16 June 2006; available at *http://trs-new.jpl.nasa.gov/dspace/handle/2014/41102* (accessed 30 August 2016).

Woods, Brian, "A Political History of NASA's Space Shuttle: The Development Years, 1972–1982," *Sociological Review* 57, no. 1 (2009): 25–46.

Yan, Haojing, Mark Dickinson, Peter R. M. Eisenhardt, Henry C. Ferguson, Norman A. Grogin, Maurizio Paolillo, Ranga-Ram Chary, Stefano Casertano, Daniel Stern, William T. Reach, Leonidas A. Moustakas, and S. Michael Fall, "High-Redshift Extremely Red Objects in the *Hubble Space Telescope* Ultra Deep Field Revealed by the GOODS Infrared Array Camera Observations," *Astrophysical Journal* 616 (2004): 63–70.

Online-Only Sources

Biography of William Ballhaus, Jr., *https://www.nasa.gov/offices/nac/members/Ballhaus-bio.html* (accessed 30 August 2016).

Biography of Burt Edelson, *http://www.arlingtoncemetery.net/biedelson.htm* (accessed 30 August 2016).

Biography of Len Fisk, *http://aoss.engin.umich.edu/files/cv/Fisk_CV.pdf* (accessed 30 August 2016).

Biography of John Mather, *http://nobelprize.org/nobel_prizes/physics/laureates/2006/mather-autobio.html* (accessed 30 August 2016).

Biography of Matt Mountain, *https://en.wikipedia.org/wiki/Matt_Mountain* (accessed 30 August 2016).

Biography of Nancy Grace Roman, Women@NASA Web Chat, 4 November 1997, *http://quest.arc.nasa.gov/women/archive/nr.html* (accessed 31 May 2010); .pdf is available in the monograph archive in the NASA Headquarters Historical Reference Collection.

Biography of NASA and NACA Center Directors, *http://www.hq.nasa.gov/office/pao/History/director.html* (accessed 30 August 2016).

Byurakan Astrophysical Observatory, *http://www.bao.am/history.htm* (accessed 30 August 2016).

Catalog of Earth Satellite Orbits, *http://earthobservatory.nasa.gov/Features/OrbitsCatalog/printall.php* (accessed 30 August 2016).

Davies, John K., "It Will Never Work! An Idea That Changed Infrared Astronomy from Space," *The Space Review*, 21 August 2006, *http://www.thespacereview.com/article/688/1* (accessed 30 August 2016).

Frommert, Hartmut, Christine Kronberg, Guy McArthur, and Mark Elowitz, SEDS, University of Arizona Chapter, *SEDS Messier Catalog* (Tucson, AZ: University of Arizona, 1994–2016), *http://messier.seds.org/* (accessed 30 August 2016).

Missions: Chandra (AXAF) at Harvard, *http://chandra.harvard.edu/* (accessed 30 August 2016).

Missions: Explorer Program at JPL, *http://www.jpl.nasa.gov/missions* (accessed 30 August 2016).

Missions: Explorer-11 Program at Goddard, *http://heasarc.gsfc.nasa.gov/docs/heasarc/missions/explorer11.html* (accessed 30 August 2016).

Missions: Explorer Series of Spacecraft, *http://history.nasa.gov/explorer.html* (accessed 30 August 2016).

Missions: Hubble Essentials, Quick Facts, *http://hubblesite.org/the_telescope/hubble_essentials/quick_facts.php* (accessed 30 August 2016).

Missions: IRAS contributors, *http://irsa.ipac.caltech.edu/IRASdocs/exp.sup/ch13/A.html* (accessed 30 August 2016).

Missions: SIRTF "Aliveness Test" Image, *http://photojournal.jpl.nasa.gov/catalog/PIA04724* (accessed 30 August 2016).

Missions: Spitzer Fact Sheet, doc. #PM 12-12-03, JPL Project Office, available at *http://www.spitzer.caltech.edu/file/97-Fact-Sheet* (accessed 30 August 2016).

Missions: Spitzer Current Location, *http://www.spitzer.caltech.edu/mission/where_is_spitzer* (accessed 30 August 2016).

Missions: Spitzer Legacy Programs, *http://irsa.ipac. caltech.edu/data/SPITZER/docs/spitzermission/ observingprograms/legacy/* (accessed 30 August 2016).

Missions: Spitzer Publication List, *http://irsa.ipac. caltech.edu/data/SPITZER/docs/spitzermission/ missionoverview/papers/* (accessed 30 August 2016).

Missions: *Spitzer Space Telescope Handbook*, ed. Schuyler Van Dyk, Michael Werner, and Nancy Silbermann, ver. 2.1, available at *http://irsa.ipac. caltech.edu/data/SPITZER/docs/spitzermission/ missionoverview/spitzertelescopehandbook/* (accessed 30 August 2016).

Raven Industries, *http://ravenind.com/about/our-history/* (accessed 30 August 2016).

White Sands Missile Range Chronology, *http:// www.wsmr.army.mil/PAO/WSHist/Pages/ ChronologyCowboystoV2stotheSpaceShuttletolasers. aspx* (accessed 30 August 2016).

Theses and Dissertations

Garber, Stephen J., "Birds of a Feather? How Politics and Culture Affected the Designs of the U.S. Space Shuttle and the Soviet Buran" (M.A. thesis, Virginia Polytechnic Institute and State University, Falls Church, VA, 2002).

Greenstein, Jesse L., "Studies of Interstellar Absorption" (Ph.D. dissertation, Harvard University, 1937).

Rieke, George H., "A Search for Cosmic Sources of 10 Exp 11 to 10 Exp 14 EV Gamma-Rays" (Ph.D. dissertation, Harvard University, 1969), American Doctoral Dissertations (X1969, p. 0236).

Rottner, Renee M., "The Life of a Project: Accomplishing Legitimacy and Mobilizing Resources for Sustained Innovation" (Ph.D. dissertation, University of California, Irvine, 2011).

Project Briefings and Press Releases

NASA ARC, "Director's memo AO-15, 31 January 1980," PP05.04, Larry A. Manning Papers 1967–1988, Box 2: Folder 3, NASA Ames History Office, NASA Ames Research Center, Moffett Field, CA.

NASA ARC, "Larry Manning, Phase A Statement of Work (SOW), presentation dated 4 October 1979," NASA Ames History Office, NASA Ames Research Center, Moffett Field, CA, AFS1070.8A, Archives Reference Collection, FC5:D4, Space Infrared Telescope Facility, Statement of Work Specification, Phase A.

NASA ARC, "SIRTF System Summary Review, presentation dated 24 July 1981," Ames Research Center SPS-14; to be deposited in NASA HRC.

NASA ARC, "Space Science Division Strategic Plans, 1980s," AFS1070.8A, Archives Reference Collection, FC5:D4, p. 5, NASA Ames History Office, NASA Ames Research Center, Moffett Field, CA; to be deposited in NASA HRC.

NASA JPL, "James Evans, Project Plan Briefing to NASA (Headquarters), presentation to George Newton, OSSA Program Manager, June 1993"; to be deposited in NASA HRC.

NASA OSSA, "Announcement of Opportunity: Shuttle Infrared Telescope Facility (SIRTF)," NASA OSSA-1-83, 13 May 1983; to be deposited in NASA HRC.

NASA OSSA, "Dick Spehalski, Office of Space Science and Applications (OSSA) Briefing, presentation dated 22 March 1990"; to be deposited in NASA HRC.

NASA OSSA, "Michael Werner, SIRTF Reconfiguration, presentation to Dr. Richard M. Obermann, science advisor, House Subcommittee on Space and Aeronautics dated 10 December 1993"; to be deposited in NASA HRC.

NASA OSSA, "Nancy G. Roman and James E. Kupperian, Orbiting Astronomical Observatories Project Briefing, presentation dated 1 December 1959," NASA TM-X-50191;

a full transcript of this briefing is available online at *http://archive.org/stream/nasa_techdoc_19630040885/19630040885_djvu.txt* (accessed 30 August 2016).

NASA OSSA, "Statement of Work (SOW), Phase A, Space Infrared Telescope Facility," AFS1070.8A, Archives Reference Collection, FC5:D4, History Office, NASA Ames Research Center, Moffett Field, CA.

Press Release, "First Successful Corona Remote Sensing Satellite Built by Lockheed Martin Marks 50-Year Anniversary," Lockheed Martin, 25 August 2010, *http://www.lockheedmartin.com/us/news/press-releases/2010/august/FirstSuccessfulCoronaRemo.html* (accessed 30 August 2016).

Press Release, "NASA's Hubble Reveals Thousands of Orion Nebula Stars," NASA press release no. 06-007, 11 January 2006, *http://www.nasa.gov/home/hqnews/2006/jan/HQ_06007_HST_AAS.html* (accessed 30 August 2016).

Press Release, "NASA Starts Work on New Space Infrared Telescope Facility," NASA JPL press release no. 98-028, 25 March 1998, *http://www.jpl.nasa.gov/releases/98/sirtfgo.html* (accessed 30 August 2016).

Press Release, "NASA's Spitzer Begins Warm Mission," SIRTF Science Center press release no. 436, 15 May 2009, *http://www.spitzer.caltech.edu/news/436-ssc2009-12-NASA-s-Spitzer-Begins-Warm-Mission* (accessed 30 August 2016).

Press Release, "New JPL Director Announces Lab Reorganization," NASA JPL press release, 2 May 2001, *http://www.jpl.nasa.gov/news/releases/2001/2001reorg.html* (accessed 30 August 2016).

Press Release, "Space Infrared Telescope Facility Investigators Selected by NASA," NASA Ames Press Release no. 84-32 (no date, c. June 1984); to be deposited in NASA HRC.

Press Release, "Space Infrared Telescope Facility Lifts Off Aboard Delta II Rocket," NASA Kennedy Space Center press release no. 76-03, 25 August 2003, *http://www.nasa.gov/centers/kennedy/news/releases/2003/release-20030825.html* (accessed 30 August 2016).

Press Release, "Space Infrared Telescope Facility Mission Status," SIRTF Science Center press release no. 146, 3 August 2003, *http://www.spitzer.caltech.edu/news/146-ssc2003-03-Space-Infrared-Telescope-Facility-Mission-Status* (accessed 30 August 2016).

Meeting Minutes

Meeting Minutes of the SIRTF Science Working Group, 12–14 September 1984.

Meeting Minutes of the SIRTF Science Working Group, 18–19 October 1984.

Meeting Minutes of the SIRTF Science Working Group, 17–19 December 1984.

Meeting Minutes of the SIRTF Science Working Group, 22 March 1985.

Meeting Minutes of the SIRTF Science Working Group, 21–22 May 1985.

Meeting Minutes of the SIRTF Science Working Group, 12 August 1985.

Meeting Minutes of the SIRTF Science Working Group, 31 October–1 November 1985.

Meeting Minutes of the SIRTF Science Working Group, 10–12 March 1986.

Meeting Minutes of the SIRTF Science Working Group, 12–14 August 1986.

Meeting Minutes of the SIRTF Science Working Group, 15–17 December 1986.

Meeting Minutes of the SIRTF Science Working Group, 1–3 June 1987.

Meeting Minutes of the SIRTF Science Working Group, 24–25 September 1987.

Meeting Minutes of the SIRTF Science Working Group, 7 December 1987.

Meeting Minutes of the SIRTF Science Working Group, 11–13 May 1988.

Meeting Minutes of the SIRTF Science Working Group, 17–19 October 1988.

Meeting Minutes of the SIRTF Science Working Group, 10–12 July 1989.
Meeting Minutes of the SIRTF Science Working Group, 14 September 1989.
Meeting Minutes of the SIRTF Science Working Group, 14–16 May 1990.
Meeting Minutes of the SIRTF Science Working Group, 11–13 September 1990.
Meeting Minutes of the SIRTF Science Working Group, 11–13 December 1990.
Meeting Minutes of the SIRTF Science Working Group, 9–11 April 1991.
Meeting Minutes of the SIRTF Science Working Group, 16–18 March 1992.
Meeting Minutes of the SIRTF Science Working Group, 29 June–1 July 1992.
Meeting Minutes of the SIRTF Science Working Group, 24–25 August 1992.
Meeting Minutes of the SIRTF Science Working Group, 1–2 December 1992.
Meeting Minutes of the SIRTF Science Working Group, 22–23 April 1993.
Meeting Minutes of the SIRTF Science Working Group, 16–17 December 1993.
Meeting Minutes of the SIRTF Science Working Group, 5–6 May 1994.
Meeting Minutes of the SIRTF Science Working Group, 8–10 September 1994.
Meeting Minutes of the SIRTF Science Working Group, 28–29 November 1994 (copy not available).
Meeting Minutes of the SIRTF Science Working Group, 9–10 May 1995.
Meeting Minutes of the SIRTF Science Working Group, 12–13 September 1995.
Meeting Minutes of the SIRTF Science Working Group, 9–10 January 1996.
Meeting Minutes of the SIRTF Science Working Group, 8–10 May 1996.
Meeting Minutes of the SIRTF Science Working Group, 9–10 September 1996.
Meeting Minutes of the SIRTF Science Working Group, 16–17 December 1996.
Meeting Minutes of the SIRTF Science Working Group, 29 April–1 May 1997.
Meeting Minutes of the SIRTF Science Working Group, 16–17 October 1997.
Meeting Minutes of the SIRTF Science Working Group, 21–23 April 1998.
Meeting Minutes of the SIRTF Science Working Group, 11–12 November 1998.
Meeting Minutes of the SIRTF Science Working Group, 6–7 April 1999.
Meeting Minutes of the SIRTF Science Working Group, 19–21 October 1999.
Meeting Minutes of the SIRTF Science Working Group, 25–26 September 2000.
Meeting Minutes of the SIRTF Science Working Group, 15–16 August 2001.
Meeting Minutes of the SIRTF Science Working Group, 10–11 May 2002.
Meeting Minutes of the SIRTF Science Working Group, 24 October 2002.

Interviews

Becklin, Eric, interview by author, Long Beach, CA, 5 January 2009 [first cited in chapter 5, note 3].
Boggess, Nancy, interview by author, Boulder, CO, 19 March 2009 [ch4n4].
Boggess, Nancy, telephone interview by author, 2 March 2009 [ch5n1].
Caroff, Larry, interview, 3 September 2008—see Manning entry.
Fazio, Giovanni G., interview by author, Cambridge, MA, 26 May 2009 [ch2].
Fisk, Lennard A., interview by Rebecca Wright, Ann Arbor, MI, 8 September 2010, *http://www.jsc.nasa.gov/history/oral_histories/NASA_HQ/Administrators/FiskLA/FiskLA_9-9-10.htm* (accessed 30 August 2016) [ch5n46].
Gallagher, David B., interview by author, Pasadena, CA, 3 March 2009 [ch8n34].
Gehrz, Robert D., interview by author, Long Beach, CA, 5 January 2009 [ch4n18].

Gehrz, Robert D., telephone interview by author, 20 January 2009 [ch5n3].

Greenstein, Jesse L., interview by Spencer R. Weart, Pasadena, CA, 7 April 1977 [Session I], Niels Bohr Library & Archives, American Institute of Physics, College Park, MD, *https://www.aip.org/history-programs/niels-bohr-library/oral-histories/4643-1* (accessed 30 August 2016) [ch1].

Greenstein, Jesse L., interview, by Spencer R. Weart, Pasadena, CA, 21 July 1977 [Session II], Niels Bohr Library & Archives, American Institute of Physics, College Park, MD, *https://www.aip.org/history-programs/niels-bohr-library/oral-histories/4643-2.html* (accessed 30 August 2016) [ch1].

Greenstein, Jesse L., interview, by Spencer R. Weart, Pasadena, CA, 26 July 1977 [Session III], Niels Bohr Library & Archives, American Institute of Physics, College Park, *https://www.aip.org/history-programs/niels-bohr-library/oral-histories/4643-3.html* (accessed 30 August 2016) [ch1].

Greenstein, Jesse L., interview by Rachel Prud'homme, Pasadena, CA, 25 February–23 March 1982, Oral History Project, California Institute of Technology Archives, *http://oralhistories.library.caltech.edu/51/01/OH_Greenstein_J.pdf* (accessed 30 August 2016) [ch1].

Harwit, Martin O., interview by David DeVorkin, Washington, DC, 20 June 1983 [Session II], Niels Bohr Library & Archives, American Institute of Physics (AIP), College Park, MD, *https://www.aip.org/history-programs/niels-bohr-library/oral-histories/28169-1* (accessed 30 August 2016) [ch2n43].

Harwit, Martin O., interview by author, Cambridge, MA, 26 May 2009 [ch2n44].

Houck, James R., interview by author, Ithaca, NY, 25 May 2009 [ch3n1].

Irace, William R., interview by author, Pasadena, CA, 18 February 2009 [ch8n42].

Kelly. Timothy J., interview by author, Boulder, CO, 20 March 2009 [ch8n17].

Kwok, Johnny, interview by author (Session I), Pasadena, CA, 26 February 2009 [ch7n5].

Kwok, Johnny, interview (Session II), 25 March 2009 [ch7n6].

Lawrence, Charles R., interview by author, Pasadena, CA, 19 January 2009 [ch9n18].

Leighton, Robert B., interview by David H. DeVorkin, Pasadena, CA, 29 July 1977 [Session I], Niels Bohr Library & Archives, American Institute of Physics, College Park, MD, *https://www.aip.org/history-programs/niels-bohr-library/oral-histories/4738-1* (accessed 30 August 2016) [ch3n5].

Manning, Larry A., and Larry J. Caroff, joint interview by author, Mountain View, CA, 3 September 2008 [ch6n22].

Martin, Frank, telephone interview, 27 March 2009 [ch4n30 and ch4nn].

McCreight, Craig, interview by author, Mountain View, CA, 2 September 2008 [ch5n26].

Neugebauer, Gerry, interview by David H. DeVorkin, Pasadena, CA, 12 August 1982, Oral History Program, National Air and Space Museum, Washington DC [ch3n5].

Pellerin, Charles J., interview by author, Boulder, CO, 19 March 2009 [ch4n38].

Rieke, George H., interview by author, Pasadena, CA, 9 June 2009 [ch2n51].

Rieke, Marcia J., interview by author, Long Beach, CA, 6 January 2009 [ch6n48].

Roman, Nancy Grace, interview by Rebecca Wright, Chevy Chase, MD, 15 September 2000, file 3636, NASA History Program Office, NASA, Washington, DC; also online at *http://www.jsc.nasa.gov/history/oral_histories/NASA_HQ/Herstory/RomanNG/RomanNG_9-15-00.htm* (accessed 30 August 2016) [ch1].

Roman, Nancy G., interview by David DeVorkin, Washington, DC, 19 August 1980, Niels Bohr Library & Archives, American Institute of Physics, College Park, MD, *https://www.aip.org/history-programs/niels-bohr-library/oral-histories/4846* (accessed 30 August 2016) [ch1].

Roman, Nancy, interview [unattributed, undated], NASA Solar System Exploration Web site, *http://*

solarsystem.nasa.gov/people/romann (accessed 30 August 2016) [ch1].

Simmons, Larry, interview by author (Session I), Pasadena, CA, 18 February 2009 [ch6n53].

Simmons, Larry, interview (Session II), 9 March 2009 [ch8n48].

Spitzer, Lyman, interview by David H. DeVorkin, Pasadena, CA, 8 April 1977 (Session 1), Niels Bohr Library & Archives, American Institute of Physics, College Park, MD, *https://www.aip.org/history-programs/niels-bohr-library/oral-histories/4901-1* (accessed 30 August 2016) [ch3n12].

Thronson, Harley A., interview by author, Long Beach, CA, 4 January 2009 [ch7n20].

Weedman, Daniel W., interview by author, Washington, DC, 27 May 2009 [ch7n13].

Werner, Michael W., interview by Sara Lippincott, Pasadena, CA, 25 July 2008, Oral History Project, California Institute of Technology Archives, *http://oralhistories.library.caltech.edu/163/* (accessed 30 August 2016) [ch5n16].

Werner, Michael W., interview by author (Session I), Pasadena, CA, 15 December 2008 [ch6n37].

Werner, Michael, interview (Session II), 19 January 2009 [ch8n28].

Westphal, James, interview by David H. DeVorkin, Pasadena, CA, 9 August 1982 (Session I), Niels Bohr Library & Archives, American Institute of Physics, College Park, MD, *https://www.aip.org/history-programs/niels-bohr-library/oral-histories/24985-1* (accessed 30 August 2016) [ch2].

Witteborn, Fred C., interview by author, Mountain View, CA, 2 September 2008 [ch3n18].

The NASA History Series

Reference Works, NASA SP-4000

Grimwood, James M. *Project Mercury: A Chronology.* NASA SP-4001, 1963.

Grimwood, James M., and Barton C. Hacker, with Peter J. Vorzimmer. *Project Gemini Technology and Operations: A Chronology.* NASA SP-4002, 1969.

Link, Mae Mills. *Space Medicine in Project Mercury.* NASA SP-4003, 1965.

Astronautics and Aeronautics, 1963: Chronology of Science, Technology, and Policy. NASA SP-4004, 1964.

Astronautics and Aeronautics, 1964: Chronology of Science, Technology, and Policy. NASA SP-4005, 1965.

Astronautics and Aeronautics, 1965: Chronology of Science, Technology, and Policy. NASA SP-4006, 1966.

Astronautics and Aeronautics, 1966: Chronology of Science, Technology, and Policy. NASA SP-4007, 1967.

Astronautics and Aeronautics, 1967: Chronology of Science, Technology, and Policy. NASA SP-4008, 1968.

Ertel, Ivan D., and Mary Louise Morse. *The Apollo Spacecraft: A Chronology, Volume I, Through November 7, 1962.* NASA SP-4009, 1969.

Morse, Mary Louise, and Jean Kernahan Bays. *The Apollo Spacecraft: A Chronology, Volume II, November 8, 1962–September 30, 1964.* NASA SP-4009, 1973.

Brooks, Courtney G., and Ivan D. Ertel. *The Apollo Spacecraft: A Chronology, Volume III, October 1, 1964–January 20, 1966.* NASA SP-4009, 1973.

Ertel, Ivan D., and Roland W. Newkirk, with Courtney G. Brooks. *The Apollo Spacecraft: A Chronology, Volume IV, January 21, 1966–July 13, 1974.* NASA SP-4009, 1978.

Astronautics and Aeronautics, 1968: Chronology of Science, Technology, and Policy. NASA SP-4010, 1969.

Newkirk, Roland W., and Ivan D. Ertel, with Courtney G. Brooks. *Skylab: A Chronology.* NASA SP-4011, 1977.

Van Nimmen, Jane, and Leonard C. Bruno, with Robert L. Rosholt. *NASA Historical Data Book, Volume I: NASA Resources, 1958–1968.* NASA SP-4012, 1976; rep. ed. 1988.

Ezell, Linda Neuman. *NASA Historical Data Book, Volume II: Programs and Projects, 1958–1968.* NASA SP-4012, 1988.

Ezell, Linda Neuman. *NASA Historical Data Book, Volume III: Programs and Projects, 1969–1978.* NASA SP-4012, 1988.

Gawdiak, Ihor, with Helen Fedor. *NASA Historical Data Book, Volume IV: NASA Resources, 1969–1978.* NASA SP-4012, 1994.

Rumerman, Judy A. *NASA Historical Data Book, Volume V: NASA Launch Systems, Space Transportation, Human Spaceflight, and Space Science, 1979–1988.* NASA SP-4012, 1999.

Rumerman, Judy A. *NASA Historical Data Book, Volume VI: NASA Space Applications, Aeronautics and Space Research and Technology, Tracking and Data Acquisition/Support Operations, Commercial Programs, and Resources, 1979–1988.* NASA SP-4012, 1999.

Rumerman, Judy A. *NASA Historical Data Book, Volume VII: NASA Launch Systems, Space Transportation, Human Spaceflight, and Space Science, 1989–1998*. NASA SP-2009-4012, 2009.

Rumerman, Judy A. *NASA Historical Data Book, Volume VIII: NASA Earth Science and Space Applications, Aeronautics, Technology, and Exploration, Tracking and Data Acquisition/Space Operations, Facilities and Resources, 1989–1998*. NASA SP-2012-4012, 2012.

No SP-4013.

Astronautics and Aeronautics, 1969: Chronology of Science, Technology, and Policy. NASA SP-4014, 1970.

Astronautics and Aeronautics, 1970: Chronology of Science, Technology, and Policy. NASA SP-4015, 1972.

Astronautics and Aeronautics, 1971: Chronology of Science, Technology, and Policy. NASA SP-4016, 1972.

Astronautics and Aeronautics, 1972: Chronology of Science, Technology, and Policy. NASA SP-4017, 1974.

Astronautics and Aeronautics, 1973: Chronology of Science, Technology, and Policy. NASA SP-4018, 1975.

Astronautics and Aeronautics, 1974: Chronology of Science, Technology, and Policy. NASA SP-4019, 1977.

Astronautics and Aeronautics, 1975: Chronology of Science, Technology, and Policy. NASA SP-4020, 1979.

Astronautics and Aeronautics, 1976: Chronology of Science, Technology, and Policy. NASA SP-4021, 1984.

Astronautics and Aeronautics, 1977: Chronology of Science, Technology, and Policy. NASA SP-4022, 1986.

Astronautics and Aeronautics, 1978: Chronology of Science, Technology, and Policy. NASA SP-4023, 1986.

Astronautics and Aeronautics, 1979–1984: Chronology of Science, Technology, and Policy. NASA SP-4024, 1988.

Astronautics and Aeronautics, 1985: Chronology of Science, Technology, and Policy. NASA SP-4025, 1990.

Noordung, Hermann. *The Problem of Space Travel: The Rocket Motor*. Edited by Ernst Stuhlinger and J. D. Hunley, with Jennifer Garland. NASA SP-4026, 1995.

Gawdiak, Ihor Y., Ramon J. Miro, and Sam Stueland. *Astronautics and Aeronautics, 1986–1990: A Chronology*. NASA SP-4027, 1997.

Gawdiak, Ihor Y., and Charles Shetland. *Astronautics and Aeronautics, 1991–1995: A Chronology*. NASA SP-2000-4028, 2000.

Orloff, Richard W. *Apollo by the Numbers: A Statistical Reference*. NASA SP-2000-4029, 2000.

Lewis, Marieke, and Ryan Swanson. *Astronautics and Aeronautics: A Chronology, 1996–2000*. NASA SP-2009-4030, 2009.

Ivey, William Noel, and Marieke Lewis. *Astronautics and Aeronautics: A Chronology, 2001–2005*. NASA SP-2010-4031, 2010.

Buchalter, Alice R., and William Noel Ivey. *Astronautics and Aeronautics: A Chronology, 2006*. NASA SP-2011-4032, 2010.

Lewis, Marieke. *Astronautics and Aeronautics: A Chronology, 2007*. NASA SP-2011-4033, 2011.

Lewis, Marieke. *Astronautics and Aeronautics: A Chronology, 2008*. NASA SP-2012-4034, 2012.

Lewis, Marieke. *Astronautics and Aeronautics: A Chronology, 2009*. NASA SP-2012-4035, 2012.

Flattery, Meaghan. *Astronautics and Aeronautics: A Chronology*, 2010. NASA SP-2013-4037, 2014.

Management Histories, NASA SP-4100

Rosholt, Robert L. *An Administrative History of NASA, 1958–1963*. NASA SP-4101, 1966.

Levine, Arnold S. *Managing NASA in the Apollo Era*. NASA SP-4102, 1982.

Roland, Alex. *Model Research: The National Advisory Committee for Aeronautics, 1915–1958.* NASA SP-4103, 1985.

Fries, Sylvia D. *NASA Engineers and the Age of Apollo.* NASA SP-4104, 1992.

Glennan, T. Keith. *The Birth of NASA: The Diary of T. Keith Glennan.* Edited by J. D. Hunley. NASA SP-4105, 1993.

Seamans, Robert C. *Aiming at Targets: The Autobiography of Robert C. Seamans.* NASA SP-4106, 1996.

Garber, Stephen J., ed. *Looking Backward, Looking Forward: Forty Years of Human Spaceflight Symposium.* NASA SP-2002-4107, 2002.

Mallick, Donald L., with Peter W. Merlin. *The Smell of Kerosene: A Test Pilot's Odyssey.* NASA SP-4108, 2003.

Iliff, Kenneth W., and Curtis L. Peebles. *From Runway to Orbit: Reflections of a NASA Engineer.* NASA SP-2004-4109, 2004.

Chertok, Boris. *Rockets and People, Volume I.* NASA SP-2005-4110, 2005.

Chertok, Boris. *Rockets and People: Creating a Rocket Industry, Volume II.* NASA SP-2006-4110, 2006.

Chertok, Boris. *Rockets and People: Hot Days of the Cold War, Volume III.* NASA SP-2009-4110, 2009.

Chertok, Boris. *Rockets and People: The Moon Race, Volume IV.* NASA SP-2011-4110, 2011.

Laufer, Alexander, Todd Post, and Edward Hoffman. *Shared Voyage: Learning and Unlearning from Remarkable Projects.* NASA SP-2005-4111, 2005.

Dawson, Virginia P., and Mark D. Bowles. *Realizing the Dream of Flight: Biographical Essays in Honor of the Centennial of Flight, 1903–2003.* NASA SP-2005-4112, 2005.

Mudgway, Douglas J. *William H. Pickering: America's Deep Space Pioneer.* NASA SP-2008-4113, 2008.

Wright, Rebecca, Sandra Johnson, and Steven J. Dick. *NASA at 50: Interviews with NASA's Senior Leadership.* NASA SP-2012-4114, 2012.

Project Histories, NASA SP-4200

Swenson, Loyd S., Jr., James M. Grimwood, and Charles C. Alexander. *This New Ocean: A History of Project Mercury.* NASA SP-4201, 1966; rep. ed. 1999.

Green, Constance McLaughlin, and Milton Lomask. *Vanguard: A History.* NASA SP-4202, 1970; rep. ed. Smithsonian Institution Press, 1971.

Hacker, Barton C., and James M. Grimwood. *On the Shoulders of Titans: A History of Project Gemini.* NASA SP-4203, 1977; rep. ed. 2002.

Benson, Charles D., and William Barnaby Faherty. *Moonport: A History of Apollo Launch Facilities and Operations.* NASA SP-4204, 1978.

Brooks, Courtney G., James M. Grimwood, and Loyd S. Swenson, Jr. *Chariots for Apollo: A History of Manned Lunar Spacecraft.* NASA SP-4205, 1979.

Bilstein, Roger E. *Stages to Saturn: A Technological History of the Apollo/Saturn Launch Vehicles.* NASA SP-4206, 1980 and 1996.

No SP-4207.

Compton, W. David, and Charles D. Benson. *Living and Working in Space: A History of Skylab.* NASA SP-4208, 1983.

Ezell, Edward Clinton, and Linda Neuman Ezell. *The Partnership: A History of the Apollo-Soyuz Test Project.* NASA SP-4209, 1978.

Hall, R. Cargill. *Lunar Impact: A History of Project Ranger.* NASA SP-4210, 1977.

Newell, Homer E. *Beyond the Atmosphere: Early Years of Space Science.* NASA SP-4211, 1980.

Ezell, Edward Clinton, and Linda Neuman Ezell. *On Mars: Exploration of the Red Planet, 1958–1978.* NASA SP-4212, 1984.

Pitts, John A. *The Human Factor: Biomedicine in the Manned Space Program to 1980.* NASA SP-4213, 1985.

Compton, W. David. *Where No Man Has Gone Before: A History of Apollo Lunar Exploration Missions.* NASA SP-4214, 1989.

Naugle, John E. *First Among Equals: The Selection of NASA Space Science Experiments*. NASA SP-4215, 1991.

Wallace, Lane E. *Airborne Trailblazer: Two Decades with NASA Langley's 737 Flying Laboratory*. NASA SP-4216, 1994.

Butrica, Andrew J., ed. *Beyond the Ionosphere: Fifty Years of Satellite Communications*. NASA SP-4217, 1997.

Butrica, Andrew J. *To See the Unseen: A History of Planetary Radar Astronomy*. NASA SP-4218, 1996.

Mack, Pamela E., ed. *From Engineering Science to Big Science: The NACA and NASA Collier Trophy Research Project Winners*. NASA SP-4219, 1998.

Reed, R. Dale. *Wingless Flight: The Lifting Body Story*. NASA SP-4220, 1998.

Heppenheimer, T. A. *The Space Shuttle Decision: NASA's Search for a Reusable Space Vehicle*. NASA SP-4221, 1999.

Hunley, J. D., ed. *Toward Mach 2: The Douglas D-558 Program*. NASA SP-4222, 1999.

Swanson, Glen E., ed. *"Before This Decade Is Out..." Personal Reflections on the Apollo Program*. NASA SP-4223, 1999.

Tomayko, James E. *Computers Take Flight: A History of NASA's Pioneering Digital Fly-By-Wire Project*. NASA SP-4224, 2000.

Morgan, Clay. *Shuttle-Mir: The United States and Russia Share History's Highest Stage*. NASA SP-2001-4225, 2001.

Leary, William M. *"We Freeze to Please": A History of NASA's Icing Research Tunnel and the Quest for Safety*. NASA SP-2002-4226, 2002.

Mudgway, Douglas J. *Uplink-Downlink: A History of the Deep Space Network, 1957–1997*. NASA SP-2001-4227, 2001.

No SP-4228 or SP-4229.

Dawson, Virginia P., and Mark D. Bowles. *Taming Liquid Hydrogen: The Centaur Upper Stage Rocket, 1958–2002*. NASA SP-2004-4230, 2004.

Meltzer, Michael. *Mission to Jupiter: A History of the Galileo Project*. NASA SP-2007-4231, 2007.

Heppenheimer, T. A. *Facing the Heat Barrier: A History of Hypersonics*. NASA SP-2007-4232, 2007.

Tsiao, Sunny. *"Read You Loud and Clear!" The Story of NASA's Spaceflight Tracking and Data Network*. NASA SP-2007-4233, 2007.

Meltzer, Michael. *When Biospheres Collide: A History of NASA's Planetary Protection Programs*. NASA SP-2011-4234, 2011.

Center Histories, NASA SP-4300

Rosenthal, Alfred. *Venture into Space: Early Years of Goddard Space Flight Center*. NASA SP-4301, 1985.

Hartman, Edwin P. *Adventures in Research: A History of Ames Research Center, 1940–1965*. NASA SP-4302, 1970.

Hallion, Richard P. *On the Frontier: Flight Research at Dryden, 1946–1981*. NASA SP-4303, 1984.

Muenger, Elizabeth A. *Searching the Horizon: A History of Ames Research Center, 1940–1976*. NASA SP-4304, 1985.

Hansen, James R. *Engineer in Charge: A History of the Langley Aeronautical Laboratory, 1917–1958*. NASA SP-4305, 1987.

Dawson, Virginia P. *Engines and Innovation: Lewis Laboratory and American Propulsion Technology*. NASA SP-4306, 1991.

Dethloff, Henry C. *"Suddenly Tomorrow Came...": A History of the Johnson Space Center, 1957–1990*. NASA SP-4307, 1993.

Hansen, James R. *Spaceflight Revolution: NASA Langley Research Center from Sputnik to Apollo*. NASA SP-4308, 1995.

Wallace, Lane E. *Flights of Discovery: An Illustrated History of the Dryden Flight Research Center*. NASA SP-4309, 1996.

Herring, Mack R. *Way Station to Space: A History of the John C. Stennis Space Center*. NASA SP-4310, 1997.

Wallace, Harold D., Jr. *Wallops Station and the Creation of an American Space Program.* NASA SP-4311, 1997.

Wallace, Lane E. *Dreams, Hopes, Realities. NASA's Goddard Space Flight Center: The First Forty Years.* NASA SP-4312, 1999.

Dunar, Andrew J., and Stephen P. Waring. *Power to Explore: A History of Marshall Space Flight Center, 1960–1990.* NASA SP-4313, 1999.

Bugos, Glenn E. *Atmosphere of Freedom: Sixty Years at the NASA Ames Research Center.* NASA SP-2000-4314, 2000.

Bugos, Glenn E. *Atmosphere of Freedom: Seventy Years at the NASA Ames Research Center.* NASA SP-2010-4314, 2010. Revised version of NASA SP-2000-4314.

Bugos, Glenn E. *Atmosphere of Freedom: Seventy Five Years at the NASA Ames Research Center.* NASA SP-2014-4314, 2014. Revised version of NASA SP-2000-4314.

No SP-4315.

Schultz, James. *Crafting Flight: Aircraft Pioneers and the Contributions of the Men and Women of NASA Langley Research Center.* NASA SP-2003-4316, 2003.

Bowles, Mark D. *Science in Flux: NASA's Nuclear Program at Plum Brook Station, 1955–2005.* NASA SP-2006-4317, 2006.

Wallace, Lane E. *Flights of Discovery: An Illustrated History of the Dryden Flight Research Center.* NASA SP-2007-4318, 2007. Revised version of NASA SP-4309.

Arrighi, Robert S. *Revolutionary Atmosphere: The Story of the Altitude Wind Tunnel and the Space Power Chambers.* NASA SP-2010-4319, 2010.

General Histories, NASA SP-4400

Corliss, William R. *NASA Sounding Rockets, 1958–1968: A Historical Summary.* NASA SP-4401, 1971.

Wells, Helen T., Susan H. Whiteley, and Carrie Karegeannes. *Origins of NASA Names.* NASA SP-4402, 1976.

Anderson, Frank W., Jr. *Orders of Magnitude: A History of NACA and NASA, 1915–1980.* NASA SP-4403, 1981.

Sloop, John L. *Liquid Hydrogen as a Propulsion Fuel, 1945–1959.* NASA SP-4404, 1978.

Roland, Alex. *A Spacefaring People: Perspectives on Early Spaceflight.* NASA SP-4405, 1985.

Bilstein, Roger E. *Orders of Magnitude: A History of the NACA and NASA, 1915–1990.* NASA SP-4406, 1989.

Logsdon, John M., ed., with Linda J. Lear, Jannelle Warren Findley, Ray A. Williamson, and Dwayne A. Day. *Exploring the Unknown: Selected Documents in the History of the U.S. Civil Space Program, Volume I: Organizing for Exploration.* NASA SP-4407, 1995.

Logsdon, John M., ed., with Dwayne A. Day and Roger D. Launius. *Exploring the Unknown: Selected Documents in the History of the U.S. Civil Space Program, Volume II: External Relationships.* NASA SP-4407, 1996.

Logsdon, John M., ed., with Roger D. Launius, David H. Onkst, and Stephen J. Garber. *Exploring the Unknown: Selected Documents in the History of the U.S. Civil Space Program, Volume III: Using Space.* NASA SP-4407, 1998.

Logsdon, John M., ed., with Ray A. Williamson, Roger D. Launius, Russell J. Acker, Stephen J. Garber, and Jonathan L. Friedman. *Exploring the Unknown: Selected Documents in the History of the U.S. Civil Space Program, Volume IV: Accessing Space.* NASA SP-4407, 1999.

Logsdon, John M., ed., with Amy Paige Snyder, Roger D. Launius, Stephen J. Garber, and Regan Anne Newport. *Exploring the Unknown: Selected Documents in the History of the U.S. Civil Space Program, Volume V: Exploring the Cosmos.* NASA SP-2001-4407, 2001.

Logsdon, John M., ed., with Stephen J. Garber, Roger D. Launius, and Ray A. Williamson.

Exploring the Unknown: Selected Documents in the History of the U.S. Civil Space Program, Volume VI: Space and Earth Science. NASA SP-2004-4407, 2004.

Logsdon, John M., ed., with Roger D. Launius. *Exploring the Unknown: Selected Documents in the History of the U.S. Civil Space Program, Volume VII: Human Spaceflight: Projects Mercury, Gemini, and Apollo.* NASA SP-2008-4407, 2008.

Siddiqi, Asif A., *Challenge to Apollo: The Soviet Union and the Space Race, 1945–1974.* NASA SP-2000-4408, 2000.

Hansen, James R., ed. *The Wind and Beyond: Journey into the History of Aerodynamics in America, Volume 1: The Ascent of the Airplane.* NASA SP-2003-4409, 2003.

Hansen, James R., ed. *The Wind and Beyond: Journey into the History of Aerodynamics in America, Volume 2: Reinventing the Airplane.* NASA SP-2007-4409, 2007.

Hogan, Thor. *Mars Wars: The Rise and Fall of the Space Exploration Initiative.* NASA SP-2007-4410, 2007.

Vakoch, Douglas A., ed. *Psychology of Space Exploration: Contemporary Research in Historical Perspective.* NASA SP-2011-4411, 2011.

Ferguson, Robert G., *NASA's First A: Aeronautics from 1958 to 2008.* NASA SP-2012-4412, 2013.

Vakoch, Douglas A., ed. *Archaeology, Anthropology, and Interstellar Communication.* NASA SP-2013-4413, 2014.

Monographs in Aerospace History, NASA SP-4500

Launius, Roger D., and Aaron K. Gillette, comps. *Toward a History of the Space Shuttle: An Annotated Bibliography.* Monographs in Aerospace History, No. 1, 1992.

Launius, Roger D., and J. D. Hunley, comps. *An Annotated Bibliography of the Apollo Program.* Monographs in Aerospace History, No. 2, 1994.

Launius, Roger D. *Apollo: A Retrospective Analysis.* Monographs in Aerospace History, No. 3, 1994.

Hansen, James R. *Enchanted Rendezvous: John C. Houbolt and the Genesis of the Lunar-Orbit Rendezvous Concept.* Monographs in Aerospace History, No. 4, 1995.

Gorn, Michael H. *Hugh L. Dryden's Career in Aviation and Space.* Monographs in Aerospace History, No. 5, 1996.

Powers, Sheryll Goecke. *Women in Flight Research at NASA Dryden Flight Research Center from 1946 to 1995.* Monographs in Aerospace History, No. 6, 1997.

Portree, David S. F., and Robert C. Trevino. *Walking to Olympus: An EVA Chronology.* Monographs in Aerospace History, No. 7, 1997.

Logsdon, John M., moderator. *Legislative Origins of the National Aeronautics and Space Act of 1958: Proceedings of an Oral History Workshop.* Monographs in Aerospace History, No. 8, 1998.

Rumerman, Judy A., comp. *U.S. Human Spaceflight: A Record of Achievement, 1961–1998.* Monographs in Aerospace History, No. 9, 1998.

Portree, David S. F. *NASA's Origins and the Dawn of the Space Age.* Monographs in Aerospace History, No. 10, 1998.

Logsdon, John M. *Together in Orbit: The Origins of International Cooperation in the Space Station.* Monographs in Aerospace History, No. 11, 1998.

Phillips, W. Hewitt. *Journey in Aeronautical Research: A Career at NASA Langley Research Center.* Monographs in Aerospace History, No. 12, 1998.

Braslow, Albert L. *A History of Suction-Type Laminar-Flow Control with Emphasis on Flight Research.* Monographs in Aerospace History, No. 13, 1999.

Logsdon, John M., moderator. *Managing the Moon Program: Lessons Learned from Apollo.* Monographs in Aerospace History, No. 14, 1999.

Perminov, V. G. *The Difficult Road to Mars: A Brief History of Mars Exploration in the Soviet Union.* Monographs in Aerospace History, No. 15, 1999.

Tucker, Tom. *Touchdown: The Development of Propulsion Controlled Aircraft at NASA Dryden.* Monographs in Aerospace History, No. 16, 1999.

Maisel, Martin, Demo J. Giulanetti, and Daniel C. Dugan. *The History of the XV-15 Tilt Rotor Research Aircraft: From Concept to Flight.* Monographs in Aerospace History, No. 17, 2000. NASA SP-2000-4517.

Jenkins, Dennis R. *Hypersonics Before the Shuttle: A Concise History of the X-15 Research Airplane.* Monographs in Aerospace History, No. 18, 2000. NASA SP-2000-4518.

Chambers, Joseph R. *Partners in Freedom: Contributions of the Langley Research Center to U.S. Military Aircraft of the 1990s.* Monographs in Aerospace History, No. 19, 2000. NASA SP-2000-4519.

Waltman, Gene L. *Black Magic and Gremlins: Analog Flight Simulations at NASA's Flight Research Center.* Monographs in Aerospace History, No. 20, 2000. NASA SP-2000-4520.

Portree, David S. F. *Humans to Mars: Fifty Years of Mission Planning, 1950–2000.* Monographs in Aerospace History, No. 21, 2001. NASA SP-2001-4521.

Thompson, Milton O., with J. D. Hunley. *Flight Research: Problems Encountered and What They Should Teach Us.* Monographs in Aerospace History, No. 22, 2001. NASA SP-2001-4522.

Tucker, Tom. *The Eclipse Project.* Monographs in Aerospace History, No. 23, 2001. NASA SP-2001-4523.

Siddiqi, Asif A. *Deep Space Chronicle: A Chronology of Deep Space and Planetary Probes, 1958–2000.* Monographs in Aerospace History, No. 24, 2002. NASA SP-2002-4524.

Merlin, Peter W. *Mach 3+: NASA/USAF YF-12 Flight Research, 1969–1979.* Monographs in Aerospace History, No. 25, 2001. NASA SP-2001-4525.

Anderson, Seth B. *Memoirs of an Aeronautical Engineer: Flight Tests at Ames Research Center: 1940–1970.* Monographs in Aerospace History, No. 26, 2002. NASA SP-2002-4526.

Renstrom, Arthur G. *Wilbur and Orville Wright: A Bibliography Commemorating the One-Hundredth Anniversary of the First Powered Flight on December 17, 1903.* Monographs in Aerospace History, No. 27, 2002. NASA SP-2002-4527.

No monograph 28.

Chambers, Joseph R. *Concept to Reality: Contributions of the NASA Langley Research Center to U.S. Civil Aircraft of the 1990s.* Monographs in Aerospace History, No. 29, 2003. NASA SP-2003-4529.

Peebles, Curtis, ed. *The Spoken Word: Recollections of Dryden History, The Early Years.* Monographs in Aerospace History, No. 30, 2003. NASA SP-2003-4530.

Jenkins, Dennis R., Tony Landis, and Jay Miller. *American X-Vehicles: An Inventory—X-1 to X-50.* Monographs in Aerospace History, No. 31, 2003. NASA SP-2003-4531.

Renstrom, Arthur G. *Wilbur and Orville Wright: A Chronology Commemorating the One-Hundredth Anniversary of the First Powered Flight on December 17, 1903.* Monographs in Aerospace History, No. 32, 2003. NASA SP-2003-4532.

Bowles, Mark D., and Robert S. Arrighi. *NASA's Nuclear Frontier: The Plum Brook Research Reactor.* Monographs in Aerospace History, No. 33, 2004. NASA SP-2004-4533.

Wallace, Lane, and Christian Gelzer. *Nose Up: High Angle-of-Attack and Thrust Vectoring Research at NASA Dryden, 1979–2001.* Monographs in Aerospace History, No. 34, 2009. NASA SP-2009-4534.

Matranga, Gene J., C. Wayne Ottinger, Calvin R. Jarvis, and D. Christian Gelzer. *Unconventional, Contrary, and Ugly: The Lunar Landing Research Vehicle.* Monographs in Aerospace History, No. 35, 2006. NASA SP-2004-4535.

McCurdy, Howard E. *Low-Cost Innovation in Spaceflight: The History of the Near Earth Asteroid Rendezvous (NEAR) Mission.* Monographs in Aerospace History, No. 36, 2005. NASA SP-2005-4536.

Seamans, Robert C., Jr. *Project Apollo: The Tough Decisions.* Monographs in Aerospace History, No. 37, 2005. NASA SP-2005-4537.

Lambright, W. Henry. *NASA and the Environment: The Case of Ozone Depletion.* Monographs in Aerospace History, No. 38, 2005. NASA SP-2005-4538.

Chambers, Joseph R. *Innovation in Flight: Research of the NASA Langley Research Center on Revolutionary Advanced Concepts for Aeronautics.* Monographs in Aerospace History, No. 39, 2005. NASA SP-2005-4539.

Phillips, W. Hewitt. *Journey into Space Research: Continuation of a Career at NASA Langley Research Center.* Monographs in Aerospace History, No. 40, 2005. NASA SP-2005-4540.

Rumerman, Judy A., Chris Gamble, and Gabriel Okolski, comps. *U.S. Human Spaceflight: A Record of Achievement, 1961–2006.* Monographs in Aerospace History, No. 41, 2007. NASA SP-2007-4541.

Peebles, Curtis. *The Spoken Word: Recollections of Dryden History Beyond the Sky.* Monographs in Aerospace History, No. 42, 2011. NASA SP-2011-4542.

Dick, Steven J., Stephen J. Garber, and Jane H. Odom. *Research in NASA History.* Monographs in Aerospace History, No. 43, 2009. NASA SP-2009-4543.

Merlin, Peter W. *Ikhana: Unmanned Aircraft System Western States Fire Missions.* Monographs in Aerospace History, No. 44, 2009. NASA SP-2009-4544.

Fisher, Steven C., and Shamim A. Rahman. *Remembering the Giants: Apollo Rocket Propulsion Development.* Monographs in Aerospace History, No. 45, 2009. NASA SP-2009-4545.

Gelzer, Christian. *Fairing Well: From Shoebox to Bat Truck and Beyond, Aerodynamic Truck Research at NASA's Dryden Flight Research Center.* Monographs in Aerospace History, No. 46, 2011. NASA SP-2011-4546.

Arrighi, Robert. *Pursuit of Power: NASA's Propulsion Systems Laboratory No. 1 and 2.* Monographs in Aerospace History, No. 48, 2012. NASA SP-2012-4548.

Goodrich, Malinda K., Alice R. Buchalter, and Patrick M. Miller, comps. *Toward a History of the Space Shuttle: An Annotated Bibliography, Part 2 (1992–2011).* Monographs in Aerospace History, No. 49, 2012. NASA SP-2012-4549.

Ta, Julie B., and Robert C. Treviño. *Walking to Olympus: An EVA Chronology, 1997–2011*, Vol. 2. Monographs in Aerospace History, No. 50, 2016. NASA SP-2016-4550.

Gelzer, Christian. *The Spoken Word III: Recollections of Dryden History; The Shuttle Years.* Monographs in Aerospace History, No. 52, 2013. NASA SP-2013-4552.

Ross, James C. *NASA Photo One.* Monographs in Aerospace History, No. 53, 2013. NASA SP-2013-4553.

Launius, Roger D. *Historical Analogs for the Stimulation of Space Commerce.* Monographs in Aerospace History, No 54, 2014. NASA SP-2014-4554.

Buchalter, Alice R., and Patrick M. Miller, comps. *The National Advisory Committee for Aeronautics: An Annotated Bibliography.* Monographs in Aerospace History, No. 55, 2014. NASA SP-2014-4555.

Chambers, Joseph R., and Mark A. Chambers. *Emblems of Exploration: Logos of the NACA and NASA.* Monographs in Aerospace History, No. 56, 2015. NASA SP-2015-4556.

Electronic Media, NASA SP-4600

Remembering Apollo 11: The 30th Anniversary Data Archive CD-ROM. NASA SP-4601, 1999.

Remembering Apollo 11: The 35th Anniversary Data Archive CD-ROM. NASA SP-2004-4601, 2004. This is an update of the 1999 edition.

The Mission Transcript Collection: U.S. Human Spaceflight Missions from Mercury Redstone 3 to Apollo 17. NASA SP-2000-4602, 2001.

Shuttle-Mir: The United States and Russia Share History's Highest Stage. NASA SP-2001-4603, 2002.

U.S. Centennial of Flight Commission Presents Born of Dreams—Inspired by Freedom. NASA SP-2004-4604, 2004.

Of Ashes and Atoms: A Documentary on the NASA Plum Brook Reactor Facility. NASA SP-2005-4605, 2005.

Taming Liquid Hydrogen: The Centaur Upper Stage Rocket Interactive CD-ROM. NASA SP-2004-4606, 2004.

Fueling Space Exploration: The History of NASA's Rocket Engine Test Facility DVD. NASA SP-2005-4607, 2005.

Altitude Wind Tunnel at NASA Glenn Research Center: An Interactive History CD-ROM. NASA SP-2008-4608, 2008.

A Tunnel Through Time: The History of NASA's Altitude Wind Tunnel. NASA SP-2010-4609, 2010.

Conference Proceedings, NASA SP-4700

Dick, Steven J., and Keith Cowing, eds. *Risk and Exploration: Earth, Sea and the Stars.* NASA SP-2005-4701, 2005.

Dick, Steven J., and Roger D. Launius. *Critical Issues in the History of Spaceflight.* NASA SP-2006-4702, 2006.

Dick, Steven J., ed. *Remembering the Space Age: Proceedings of the 50th Anniversary Conference.* NASA SP-2008-4703, 2008.

Dick, Steven J., ed. *NASA's First 50 Years: Historical Perspectives.* NASA SP-2010-4704, 2010.

Societal Impact, NASA SP-4800

Dick, Steven J., and Roger D. Launius. *Societal Impact of Spaceflight.* NASA SP-2007-4801, 2007.

Dick, Steven J., and Mark L. Lupisella. *Cosmos and Culture: Cultural Evolution in a Cosmic Context.* NASA SP-2009-4802, 2009.

Dick, Steven J. *Historical Studies in the Societal Impact of Spaceflight.* NASA SP-2015-4803, 2015.

Index

A

absolute zero, 19–22
Advanced X-ray Astrophysics Facility (AXAF). *See also* Chandra X-ray Observatory
    advocacy for, 98
    budgets for, 78–81, 87
    defunded, 80–81
    Great Observatories program, 72–81, 92
    "New Start," 72–76, 86
    orbit of, 87–88
    priority of, 47, 73–74, 83, 87, 98
    renamed Chandra X-ray Observatory, 88
    scaled back, 88, 104
Advisory Committee on the Future of the U.S. Space Program, 84, 96
aeronautics versus astronomy, 63–68, 89–90
airborne astronomy, 14n3, 25, 29, 36–38, 41
Allen, Lew, 76–77
Alsos mission, 6
American Astronomical Society, 98
Ames Research Center (ARC), 36–38, 41–43, 106
    aeronautics versus astronomy, 64–66, 89–90
    collaboration with NASA, 49
    data communication plan, 95
    flight testing at, 90–91
    influence of, 64–66, 91, 94
    infrared astronomy panel, 46
    infrared observations, 67
    isolationism of, 64
    project management, 88–89, 93–94
    projects at, 64–66
    relationship with JPL, 95
    relationship with NASA, 63–66
    SIRTF Phase B RFP, 96
    SIRTF project management bid, 92–94
    SIRTF Space Science Division, 89–90
    SIRTF Study Office, 54, 70, 83–84, 93
    SOFIA development, 105
Anderson, Clinton, 34
Announcement of Opportunity
    defined, 50, 54
    of Explorer, 52
    of SIRTF, xiv, 50, 58–59, 61, 69, 104, 117–118
Apollo 1 fire (AS-204), 35
Apollo program, xii, 35, 41, 64, 78
Applied Physics Lab (APL) at Johns Hopkins, 7
archives, xiii, 11, 74n33
Army Ballistic Missile Agency (ABMA), 10

astronomers
    attitudes toward infrared research, 17–22, 24, 29, 31–33
    interpretation of infrared research, 27
astronomy. *See also* infrared astronomy; optical astronomy
    versus aeronautics, 63–68, 89–90
    airborne, 14n3, 25, 29, 36–38, 41
    balloon-borne observations, 13–15, 26, 36. *See also* rockoon
    balloon-borne observations (Hoffman's paper), 42
    birth of stars, 27, 74, 118
    budgets for, 23–25, 36–37
    collaboration in, xi, 130–131, 137–151
    consensus in, 53
    "Decade of the Infrared," 97
    first infrared astronomer, 20
    German contributions to, 6–8
    high-energy, 13–14, 43
    influence of, 83
    infrared, skepticism toward, 29, 33
    infrared technique studies, 49
    microwave, 17, 26n46, 33, 56
    military contributions to, 6, 8, 17–18, 29–30
    optical, influence of, 83
    optical working groups, 43
    optimal observational conditions, 24
    physicists' contribution to, 14, 20–23, 25, 29–30, 37
    planetary, 83, 110, 118, 136
    radio, 10, 17, 22, 33
    radio, funding for, 53, 56
    rivalries in, 64–66
    rocket-borne experiment costs, 25
    rocket-borne observations, 5, 8, 11, 13–14, 29, 36
    rocket-borne observation failures, 34
    solar, 36, 43, 54
    sortie-mode, 37–38, 41–43, 46–47, 69
    space-borne observations, 33–35, 136
    spectroscopy, 6–8
    and technology, 57
    ultraviolet, 8, 11, 13, 17–18, 54, 56
    x-ray, 11, 36, 43, 54
Astronomy and Astrophysics for the 1970s (report), 52
Astronomy Survey Committee, 36, 52, 55
Astronomy Working Group (Optical), 43–45, 52
Astrophysical Letters and Communications, 84
Atlas rocket, 111–112, **113**
    for SIRTF launch, 102, 104–105, 109
Atomic Energy Commission, 14
Augustine, Norman, 84, 96
Augustine Committee report, 24n44, 84, 96–98

B

Baade, Walter, 14
Bader, Michel, 37–38, 65
Bahcall, John, 97
Bahcall report, 97, 106, 110, 118
Ball Aerospace, 62, 111, 138–150
    cryogenic technology, 125–127
    Hubble Space Telescope, 109
    SIRTF Infrared Spectrograph (IRS), 125–127, 129–131
    SIRTF meetings, 139
    SIRTF Phase B subcontractor, 114, 123
    SIRTF project organization chart, **124**
Ballhaus, William, 90
balloon-borne observations, 8, 13–14, 29, 36. *See also* rockoon
    abandoned, 26
    and gamma rays, 15
    Mars research, 21
    paper on, 42
balloons, altitudes of, 8, 13
BDM Corporation, 80
Becklin, Eric
    Astronomy Survey Committee infrared panel, 36
    Becklin-Neugebauer object, 27, 29, 41
    FIRST, 51
    infrared observations, 27
    infrared astronomy panel, 68
    SIRTSAG, 49
    SIRTF early design, 41
    SIRTF near-infrared camera, 62
    Very early system design study (SIRTF), 52
Becklin-Neugebauer object, 27, 29, 41
Beggs, James M., 79
Bell Telephone Laboratories, 17
Beta Pegasus, 31
Betelgeuse, 1, 3, 19, 20n30
Bevatron, 15
Bicay, Michael, 123
Big Bang, 17, 56, 68, 110, 136
Boggess, Nancy
    COBE program scientist, 61
    FIRST, 51
    Haverford College, 66
    Infrared Astronomical Satellite (IRAS), 56–57
    NASA senior staff scientist, 43
    Science Working Group, **69**
    SIRTF Announcement of Opportunity, 54
    SIRTF program scientist, 61–63, 138
bolometers, 20–23, 67
Boynton, Paul E., 51
Brookhaven National Laboratory, 15
Broomfield meeting, 109–113
brown dwarfs, xii, 110, 118, 122
budgets
    federal, 77–78
    federal (timetable), **78**
    fiscal policy (U.S.), 77–78
    Hubble budget, 72–73, 86n8, 102
    Hubble budget shortfalls, 79, 86
    infrared astronomy, 47
    NASA budget authority, **86**
    NASA budget cuts, 57, 101
    NASA budget shortfalls, 77, 84
    SIRTF budget, 54–55, 118, 139
Burmester, William, 126
Bush, George Herbert Walker, 104
Byurakan Observatory, 10

C

California Institute of Technology (Caltech), 130, 145
    Bruce Murray at, 87
    Cassini program, 87
    Eric Becklin at, 36, 41, 68
    Gerry Neugebauer at, 27, 37, 46
    infrared astronomy panel, 46
    infrared research at, 20n 30, 20n 32, 43
    James Westphal at, 18, 33
    Jesse Greenstein at, 9, 34
    management of JPL, 92, 98
    Michael Werner at, 67–68
    Optical Astronomy Working Group, 43
    planetary astronomy, 87
    relationship with JPL, 92, 98
    Robert Leighton at, 31, 36
    Robert Wildey at, 33
    SIRTF project organization chart, **124**
    Two-Micron Sky Survey telescope, 31, **32**
    Wesley Huntress at, 104
Cambridge Research Laboratories (AFCRL), 30, 36, 38, 46, 49. *See also* military contributions to science; U.S. Navy
Cape Canaveral, 15, **135**
Capps, Richard W., 51
Caroff, Lawrence J.
    Ames theoretician, 38, 41, 65
    Infrared, Submillimeter, and Radio Astrophysics Branch (NASA), 90, 104
    return to Ames, 106
    SIRTF near-infrared camera, 62
    SIRTF program scientist, 108
    SIRTF project management bids, 93
    SIRTF/SOFIA advocacy program, 105–106, 138
    SIRTF/SOFIA competition, 106
    SIRTF sortie-mode design, 41–43
    Very early system design study (SIRTF), 52
Carter, Jimmy (James Earl), 56
Cassini program, 77n45, 87
Catalog of Nebulae and Star Clusters, 2, 29
Center for Radio Physics and Space Research, 23
Cerenkov counter, 15
Challenger (Space Shuttle), xiii, 43n3, 78–81, 85n6, 99
Chan, K.I. Roland, 51
Chandra X-ray Observatory, 57, 88. *See also* Advanced X-ray Astrophysics Facility (AXAF)

China Lake facility (U.S. Navy), 18
Clinton, Bill, 104
Cold War, xiii, 10, 78
Comet Rendezvous Asteroid Flyby (CRAF), 77n45, 87n13
comets, 2, 5, 56
Committee on Space Astronomy and Astrophysics (CSAA), 53
Compton, Dale I., 51, 90
Compton Gamma Ray Observatory (CGRO), 53, 57.
    See also gamma ray detectors
  Great Observatories program, 72–81
  priority of, 73, 75
Congress (U.S.)
  fiscal policy, xiii–xiv, 66, 77–78, 81n53
  Great Observatories program, 74, 77
  House Appropriations Committee, 104
  House Subcommittee on Space and Aeronautics, 113
  infrared telescopes (IRT), 56
  NASA appropriations, xiv, 53, 85–87, 101–102, 113, 118
  NASA oversight, xiii, 35, 88, 98n46, 100, 107–108, 124
  "New Start," 71
  party control of, 104
  SIRTF funding, 51, 53–54, 58
  Space Shuttle, 36, 47
contamination, 45, 47–50, 55
Copernicus mission, 56
Cornell University, 20, 36, 139
  Carl Sagan at, 87
  infrared spectroscopy at 20, 23–26
  IRAS team listing, 62n5
  James Houck at, 25–26, 36, 51, 62, 98, 122
  Martin Harwit at, 23–26, 66
  Michael Werner at, 66–67
  Philip Morrison at, 14
  SIRTF meetings, 139
  SIRTF project organization chart, **124**
  Thomas Soifer at, 27
Cosmic Background Explorer (COBE), 17n17, 56–58, 61, 67n17
  cryogenically cooled, 106
  "New Start," 72
  successes of, 105
cosmic background radiation (CMB), 17, 26n46, 56, 67
  infrared, 26
Cosmic Discovery: The Search, Scope, and Heritage of Astronomy, 57, 73–74, 81
cosmic rays, 8, 69
Cosmotron, 15
"Crisis in Space and Earth Science, the" report, 84
Cruikshank, Dale, 6n9, 37, 83, 88
cryogenic telescope assembly (CTA), 149
  SIRTF project organization chart, **124**
cryogenically cooled infrared telescope (IRT)
  contamination of, 55
  design for, 49
  feasibility study, 46

Hughes report, 50
  problems with, 125–127
  SIRTF designed as, 41–42, 44–45, 58
  weight of, 111
CSAA (Committee on Space Astronomy and Astrophysics), 53. *See also* Space Science Board (SSB)
Cummings, Ramona, 107

D

Davidson, Arnold, 22–23
Davidson, Kris, 27
decadal surveys
  Bahcall report, 97, 106, 110, 118
  Field report, 55–59, 66, 73, 75, 84, 97
  Greenstein report, 36, 46, 53–59, 73, 97
Deep Space Network, 132
Delta rocket, 112, **113**
  for SIRTF launch, xi, 102, 111–112, **135**
Department of Defense (DOD), 22, 34, 56, 77
Desert Storm, 97–99, 101
Design Optimization Study Team (DOST), 50
dewar, 111, 114, 119
  defined, 106
  pressurization, 125n18, 125–127
  and SIRTF MIC, 129–131
Dubin, Maurice, 43
dust
  atmospheric, 53
  contaminant, 45, 69
  interstellar, 3–5, 13, 44
Duthie, Joseph, 15

E

Earth-trailing orbit, 102, 111–115, 137, 143, 151
Edelson, Burton I., 76–77, 79
Edison program, 105–108
  passive cooled IRTs, 106, 111
  projected cost of, 107
  SIRTF/Edison competition, 107–108
Einstein, Albert, 37
Einstein program, 56
Einstein Observatory, 92
Eisenhardt, Peter, 95
Elachi, Charles, 98
electromagnetic spectrum, 16, 36, 56–57, 72
  wavelengths, **17**
energy crisis, 36, 46–47
Erickson, Edwin F., 51, 64–65
Essex Corporation, 74
European Space Agency, 46n8, 56, 88, 106, 122
Evans, James A., 99, 103, 109
Evans, Neal J., 51
Everitt, C.W.F., 37
Explorer program, 14n4, 53, 57–58, 63–64
  Announcement of Opportunity, 52
  observation of gamma rays, 15n12

F

Fairbank, William M., 37
Fanson, James, 100n53, 117, 144
"faster, cheaper, better," 101, 104, 113, 133. *See also* Goldin, Daniel
Fazio, Giovanni
  balloon-borne observations, 15–16, 26–27
  FIRST, 51
  gamma ray research, 15–16
  Infrared Astronomical Satellite (IRAS), 71
  infrared telescopes (IRT), 30, 48, 106
  Science Working Group PI, **69**, 70, 98
  SIRTF Infrared Array Camera (IRAC), 26, 62, 92, 105, **122**
  SIRTF project organization chart, **124**
federal budget, 77–78
  timetable, **78**
Field, George, 55, 74n33
Field report, 55–59, 66, 73, 75, 84, 97
First Gulf War, 97–99, 101
first infrared astronomer, 20
first orbiting observatory, 15–16, **16**
fiscal policy (U.S.), 77–78
Fisk, Lennard A.
  departure from NASA, 104
  Infrared Astronomy Mission, 99
  Office of Space Science and Applications (OSSA), **75**, 79–80, 87
  SIRTF project management bids, 92–93
  SIRTF renamed, 102
Fletcher, James C., 79–80
flight testing, 90–91
Focal-plane Instruments and Requirements Science Team (FIRST), 50–51, 53, 54
  final report, 53
focal-plane investigations, 61–62
Ford, Gerald, 79
Friedman, Herbert, 23
Friedman report, 52
Fuchs, Art, 98

G

galaxies, 14, 29, 110, 118, 122
Galileo program, 99, 105
Gallagher, David B., 100n53, 143
  personality of, 131, 138–139, 145–146
  quoted, 138–139, 146
  SIRTF project manager, 126, 130–131, 133–134, 143
gamma ray detectors, 15–16, **24**, 26–27, 30, 46. *See also* Compton Gamma Ray Observatory (CGRO)
gamma rays (cosmic), 13–17, 26, 54–56
  detected by OSO-1, 26
Gaustad, John E., 36
Gehrz, Robert D.
  advocacy for SIRTF, 138
  Bahcall report, 110
  Design Optimization Study Team (DOST), 50
  FIRST, 50, 51
  NASA RFP for facility design study, 52
  Shuttle-based SIRTF, 55–56
  SIRTF dewar, 127
  SIRTF initial design studies, 49
  SIRTF near-infrared camera, 62
Gemini program, 35
German contributions to astronomy, 6–8
Gillett, Fred C., 36, 49
  FIRST, 51
  Infrared and Radio Astrophysics Branch (NASA), 93
  Infrared Astronomy Panel, 97
  IRAS team listing, 62n5
  personality of, 97
Gilman, Dave, 73–74
Goddard Space Flight Center (GSFC), 38, 42–43, 95
  flight testing, 91
  Hubble Space Telescope, 148–149
  influence of, 64
  Infrared Array Camera (IRAC), **122**, 123
  infrared astronomy panel, 46
  and Optical Astronomy Working Group, 44
  passive cooled IRTs, 111
  project management, 89
  SIRTF construction, 126
  SIRTF project management bid, 92–94, 105
Gold, Thomas, 23
Goldin, Daniel. *See also* "faster, cheaper, better"NASA Administrator, 100–102, 104, 109, 134
  SIRTF construction, 123–124, 133–134
gravity, 8, 37, 102
Great Observatories program, 72–81, 84, 123n11, 132
  appeal of, 86
  congressional support of, 104
  Harwit promotional booklet, 74, 81
  NASA support for, 105
  project interdependence of, 87, 125
  publicity for, 74, 81, 107
Greenstein, Jesse L., **6**, **9**
  Astronomy Survey Committee chair, 36
  Caltech astronomy department, 9–10
  interstellar dust, 4–5
  rocket-borne observations, 5–8, 34
  solar spectroscopy experiment, 8
Greenstein report, 35–36, 46, 53–59, 73, 97.

H

Hafner, Everett, 15–16
Hale Observatory at Mount Palomar, 9, 14, 19, 27, 34
Hall, Freeman F., 30–31, 33
Halley's Comet, 56
Hanel, Rudolf, 46
Harper, Doyal A., 49, 51
Harvard-Smithsonian Astrophysical Observatory (SAO), 26, 122
  SIRTF project organization chart, **124**

Harvard-Smithsonian Center for Astrophysics (CfA), 20, 26–28, 62, 98
   SIRTF project organization chart, **124**
Harvey, Paul M., 51
Harwit, Martin, 52
   book, 57, 73–74, 81, 138
   Cornell, 23, 66–67
   history of infrared astronomy, 22
   ISO, 88n17
   Naval Research Laboratory (NRL), 23, 67
   rocket-borne experiments, 25, 125
   theoretical paper, 27
Haverford College, 66
Hawarden, Tim, 106–107, 111
heat detectors, 18, 30
Herschel mission, 46n8
Hertzsprung-Russell diagram, 3–4, **5**, 22–23
Hoffmann, William F., 42–43, 46, 49, 51, 74n33
Honeywell, 62
Houck, James R.
   Broomfield meeting, 109–110
   Cornell, 25, 36, 62, 98
   Design Optimization Study Team (DOST), 50
   FIRST, 51
   Infrared Astronomical Satellite (IRAS), 62
   Infrared Astronomy Panel, 97
   NASA Director of Astrophysics, 104
   personality of, 131
   rocket-borne experiments, 125
   Science Working Group PI, 69, **69**, 97–98, 109–110, 118
   SIRTF contract competition, 49
   SIRTF Infrared Spectrograph (IRS) PI, 25, 62, 97–98, 110, 118, **122**
   SIRTF IRS filter delamination, 127, 129–131
   SIRTF project organization chart, **124**
House Appropriations Committee, 104
House Subcommittee on Space and Aeronautics, 113
Hoyle, Fred, 23, 67
Hubble Space Telescope, 53, 65, 148–149
   budget for, 72–73, 86n8, 102
   budget shortfalls, 79, 86
   construction of, 91
   development of, 9
   effects of radiation on, 70
   expense of, 57, 118
   Great Observatories program, 72–81, 92
   "Hubble Syndrome," 86
   infrared detectors for, 98
   launch of, 34
   Near Infrared Camera and Multi-Object Spectrometer (NICMOS), 130
   and Optical Astronomy Working Group, 43
   orbit of, 87–88, 114
   as Shuttle payload, 45
   priority of, 53–54, 75, 85n6, 98n46, 99
   problems with, 99–101, 130
   repairs to, 72, 100, 109, 114

Hughes Aerospace, 49, 53
Hughes report, 50, 52. *See also* "Large Cooled Infrared Telescope Facility for Spacelab"
Huntress, Welsey T., 104

I

Infrared Array Camera (IRAC), xi, 26, 62, 70, 92, **122**
   Cold Assembly, **120**
   defined, 118
   SIRTF integration flow, **89**
   SIRTF project organization chart, **124**
   successes of, 136
Infrared Astronomical Satellite (IRAS), 61–62, 97
   budgets for, 72
   cost, actual, 130, 148
   cryogenically cooled, 106
   cryostat, 129–130
   defined, 53
   dewar integration, 126, 130
   as free flying satellite, 56
   launch, 117
   mission completion, 59
   orbit of, 114
   priority of, 58, 72
   program scientist (Boggess), 61
   projected cost of, 102
   results from, 59
   successes of, 88, 105
   team listing, 62n5
infrared astronomy, 1, 11, 19–21, 33, 136. *See also* astronomy; optical astronomy
   age of, xii
   at Ames Research Center, 65
   budgets for, 23–25, 36
   community, 38
   coolants, 19
   "Decade of the Infrared," 97
   defined, 46
   funding for, 53
   history of, 22, 29
   innovations, xii–xiii, 117–118, 125
   map of Venus, 33
   masers, 67
   observations, accuracy of, 27
   photography, **4**
   photometry, 20–21, 45
   and planetary astronomy, 83
   physicists' contribution to, 20–23
   preparatory research, 95–96
   research neglected, 16–18
   rewards of, 24, 27–28
   as separate field of research, 25–29, 33
   and SIRTF, 50, 54
   skepticism toward, 21–22, 29, 31–33, 36, 65
   and Space Shuttle, 46
Infrared Astronomy Panel
   Science Working Group (SWG), 35–36

Space Science Board (SSB), 46
infrared cosmic background radiation, 26
infrared detectors, **31**
    cryogenically cooled, 30–31, **31**
    devices (night vision), 6
    early, 20
    effects of radiation on, 70
    innovations, 95, 104, 117–118
    invented, 15
    lead-sulfide (PbS), 6, 18, 20, 30–31, **31**
    limits of, 24, 33–34, 70–71
    Mars research, 21
    military, 70–71
    military applications of, 18
    on SIRTF, 99
    problems with, 19, 30, 129
    and semiconductors, 18–20
    successes of, 22–23, 29–33
Infrared Processing and Analysis Center, 123–134
infrared radiation, xi, 17–18, 21, 30, 119
infrared remote sensing, 46
infrared sky surveys, 25–26, 29–33, 36
    by SIRTF, 43–44
    deep space, 110
    Infrared Astronomical Satellite (IRAS), 56–57
    international collaboration on, 53, 56
Infrared Space Observatory (ISO), 88, 106–107, 122
Infrared Spectrograph (IRS), xi, 61, **120**
    Ball Aerospace, 123
    described, 118–119
    filter delamination, 125, 127, 129–131
    Principal Investigator (Houck), 62
    program scientist (Boggess), 61
    SIRTF integration flow, **89**
    SIRTF project organization chart, **124**
Infrared Telescope Facility (IRTF), 66
infrared telescopes (IRTs), 23, 29–30, **32**, 45
    contamination of, 45, 47–49
    cryogenically cooled, 31, 36–38, **39**, 41–42
    design optimization study, 53
    Edison program, 105
    focal plane, 61–62
    funding for, 56
    multiple missions, 54
    passive cooled, 106–107, 111
    planetary astronomy, 83
    sensitivity of, 48
    Shuttle-based, 46–48, 106
    space-borne, xii, 118
    Spacelab-2 payload, 62
Institute for Theoretical Astronomy, 67
International Solar Polar Mission, 56
International Space Station (ISS), 84
Irace, William R., 126, 138–139
    quoted, 139–140, 148
    SIRTF cryostat, 130–131
    SIRTF software, 133–134
ITT Federal Laboratory, 30

J

James Webb Space Telescope (JWST), 108n24, 111n33, 122
Jennings, Richard, 46
Jet Propulsion Laboratory (JPL), 43, 76–77, 138, 151
    Astronomy and Physics Directorate, 130
    data communication plan, 95
    discretionary funding, 99
    Earth and Space Science, 98
    Edison program, 108
    employees, 98
    facilities, 141
    flight testing, 91
    Hubble Space Telescope, 148
    influence of, 64
    internal review, 96
    Mars missions, 135
    Office of Technology and Applications, 99
    partnered with Ames, 95
    personnel reassignment, 99
    planetary astronomy, 87
    project management, 89
    relationship with Caltech, 92, 98
    SIRTF mission operations, 134
    SIRTF objectives, 112
    SIRTF Phase B, 123, 132
    SIRTF Programmatic Commitment Agreement, 114
    SIRTF project management, 123, 140–142, 144
    SIRTF project management bid, 92–94
    SIRTF project organization chart, **124**
    SIRTF redesigned, 113–114
    transfer of SIRTF, 94–97, 102, 105
Johns Hopkins Applied Physics Lab (APL), 7
Johnson, Harold L., 20–23
Johnson, Rodney O., 52
Johnson Space Center (JSC), 44, 64
Jupiter, 20n30, 23, 25
Jura, Michael, 27, 63
    Science Working Group, 69, **69**, 83, 110

K

Kaplon, Mort, 15
Keck I telescope at Mauna Kea, 9
Kelly, Tim, 125–127, 138, 146–151
Kennedy, John F., 21, 78
Klarmann, Joseph, 15
Kleinmann, Douglas, 27
Kleinmann-Low nebula, 27, 29
Knacke, Roger F., 51
Kondracki, Henry, 23, 25
Kuiper, Gerard P., **6**, 22n37, 26, 83
    airborne experiments, 37
    infrared research, 5–9, 37
    Lunar and Planetary Laboratory (LPL), 21
Kuiper Airborne Observatory, 48, 67–68
Kunde, Virgil C., 51
Kupperian, James E., 14n5
Kvant-1, 80

Kwok, Johnny (John), 112, 138
    personality of, 151
    quoted, 148, 151
    SIRTF chronological changes timetable, 112
    SIRTF mission operations system manager, 134
    SIRTF organization chart, **96**
    SIRTF software, 132–133
    Trailing-Earth orbit, 102–104, 114, 132, 143

L

Langley Research Center, 36, 64
Lanzerotti, Louis, 106
LaPiana, Lia, 139
"Large Cooled Infrared Telescope Facility for Spacelab, a," 50. *See also* Hughes report
Large Space Telescope (LST). *See* Hubble Space Telescope
Larson, Harold P., 36, 51
Lawrence Berkeley Laboratory, 15
Lawrence, Charles, 123, 125, 143–145
lead-sulfide (PbS) detectors, 18, 20, 30–31, **31**
Leighton, Robert, 31, 33, 36
Lester, Daniel, 106
Lick Observatory, 44, 46
Lincoln, Abraham, 36
Lockheed Martin, 142, 144, 146, 148–149, 151
    Hubble Space Telescope, 148
    merger with Martin Marietta, 132, 141
    organizational culture, 91, 133–134
    SIRTF meetings, 139
    SIRTF Phase B subcontractor, 114, 123, 125–127
    SIRTF project organization chart, **124**
    SIRTF software, 132–134, 138,
Lockheed Martin Space Systems Company (LMSSC), 132, 134–135
Low, Frank, **39**, 111n33
    advocacy for SIRTF, 108
    airborne infrared experiments, 25
    Astronomy Survey Committee, 36–38
    bolometer measurements, 21
    FIRST, 51
    influence of, 33
    infrared observations 21–23, 33
    IRAS team listing, 62n5
    Lunar and Planetary Laboratory, 21–22
    NASA contract competition review panel, 49
    Orion observations, 27, 29
    personality of, 38, 97
    Science Working Group, **69**, 103
    SIRTF as free flying satellite, 55, 59
    SIRTF cryostat, 130–131
    SIRTF facility scientist, 62, 88, 97, 117, 138
    warm-launch concept, 111–112
Lowell Observatory, 44
Lowell, Percival, 21, 83
Lunar and Planetary Laboratory (LPL), 21–22
lunar science, 19–21, 35

M

M42 (Orion Nebula), 2–3, **2–4**
Macenka, Steve, 100n53
Manning, Larry A., 90
Mariner program, 33–34, 77n45, 83
Mark, Hans, 65
Mars
    canals, 21, 83
    infrared observations of, 23
    life on, 21, 83
    Lowell's books, 21
    mapped by Mariner program, 33
Mars Observer, 99, 105
Marshall Space Flight Center (MSFC), 43–44, 76
    Advanced X-ray Astrophysics Facility (AXAF), 74, 87, 92
    flight testing, 91
    Hubble Space Telescope, 92, 148
    influence of, 64
    and Optical Astronomy Working Group, 44
    project management, 89, 91
    SIRTF project management bid, 92–95
    Space Shuttle engine bells, 107
    Space Shuttle main engines, 87
Martin, Frank
    Infrared Astronomical Satellite (IRAS), 72
    NASA Director of Astrophysics, 55–58, 72, 132–134, 148
    project management, 55–58, 72
    quoted, 57–58, 142, 148–149
    use of Martin Harwit's book, 57, 73
Martin Marietta Aerospace
    merger with Lockheed Martin, 132, 141
    NASA RFP for facility design study, 52
    "Payload/Orbiter Contamination Control Requirement Study," 49, 52
    SIRTF initial design, 49
Martz, Dowell, 18–19
masers, 67
Massachusetts Institute of Technology (MIT), 15, 51
Mather, John C., 17n17 51, 67, 108n24
McCarthy, Steve G., 50, 52
McCreight, Craig, 70–71
McDonald Observatory, 20, 22, 44
McNutt, Douglas, 36
Mercury (planet), 33
Mercury program, 35
Meredith, Leslie, 52
Merrill, K. Michael, 51
Messier, Charles, 2–3, 5, 29
microwave astronomy, 17, 26n46, 33, 56
microwave radiation, 17
Mikulski, Barbara, 98
military contributions to science, xii, 8, 17–18, 29–30. *See also* Cambridge Research Laboratories (AFCRL); U.S. Navy
Mir space station, 80

Moffett Field, 37
Moon, infrared observations of, 19–20
Moon landings, xii, 35, 42, 64
Moorwood, Alan F.M., 51
Morrison, Philip, 14–15
Moseley, Harvey, 111, 122
Multiband Imaging Photometer for Spitzer (MIPS), xi, 26, 62, 70, **120–122**, 123, 147
   defined, 118
   failures of, 129, 131n33
   SIRTF integration flow, **89**
   SIRTF project organization chart, **124**
   wiring failure, 131n33
Multiple Instrument Chamber (MIC), 119, **120–121**, 126, 129
Murray, Bruce, 18–19, 33, 87

N

National Academy of Sciences, 35–36
   Astronomy Survey Committee, 55
   budgets for infrared astronomy, 47
   feasibility study of cryogenically cooled IRT, 51
   Field report, 54–55, 66
   Greenstein report, 46, 52
   mission priorities, 53, 56
   recommendations for major projects, 53
   "Report on Space Science," 52
   Scientific Uses of Space Shuttle (report), 52
   SIRTF study team, 54
   Space Science Board (SSB), 8–9, 17–18, 20, 33, 53
National Advisory Committee for Aeronautics (NACA), 10
National Aeronautics and Space Administration (NASA), 9, 34. *See also* Office of Space Science and Applications (OSSA)
   Advisory Council, 84
   and airborne astronomy, 38
   Ball Aerospace contract, 126
   budget authority, **86**
   budget cuts at, 57, 101
   budget shortfalls, 77, 84
   congressional support of SIRTF, 113
   construction bid process, 123–124
   contract competition, 49
   criticism of, 23, 34–35, 101
   decadal surveys. *See* Greenstein report, Field report
   departments of, 46, 72
   Edison program, 106–107
   feasibility study of cryogenically cooled IRT, 51
   founding of, 10–11, 13–14
   Great Observatories program, 57
   hiring freeze at, 64–65
   Historical Reference Collection (HRC), xiii
   Hubble budgets, 72–73
   infrared astronomy program budgets, 47–49, 58–59
   infrared panel, 35–36
   Infrared Processing and Analysis Center, 134
   and infrared research, 17–18, 21, 24, 36, 38
   Infrared, Submillimeter, and Radio Astrophysics Branch (NASA), 90
   instrument design competition, 20
   and Kuiper's research, 21
   Mars missions, 133, 135
   mission cancellations, 98n46
   mission life cycle (full), 95
   mission priorities, 72–81, 83, 98, 104–106, 117
   naming policy, 34
   "non-competition competition," 92
   Office of Aeronautics and Space Technology, 93
   Office of Space Science and Applications (OSSA), 11, 46–47, 72, 75, 79–81
   and Optical Astronomy Working Group, 44
   orbiting observatory program, 15–16
   passive cooled IRTs, 107
   perception of, xviii, 100, 108–109
   program manager, 98
   project list, **85**
   project management, xii–xiii, 61–62, 95, 101, 109, 137, 145–146
   project scientist, role of, 146
   Quality and Mission Assurance, 126
   relationship with Ames Research Center, 49, 63–66
   relationship with Caltech/JPL 92, 98
   report timeline, 52
   RFP for facility design study, 52
   Shuttle engineering studies, 48–49
   Shuttle Infrared Telescope Science Accommodation Group (SIRTSAG), 49
   Shuttle program budgets, 42–43, 72
   SIRTF Announcement of Opportunity, xiv, 50, 54
   SIRTF budget, 54–55, 118, 139
   SIRTF construction, 124
   SIRTF infrared spectrograph (IRS), 126
   SIRTF internal reviews, 123
   SIRTF "New Start," xiv, 71–72, 74
   SIRTF project organization chart, **124**
   SIRTF trailing-Earth orbit, 115
   Solar Astronomy Group, 63
   sounding rocket program, 57
   Space and Earth Sciences Advisory Committee (SESAC), 77n45, 83
   space policy, 96–97, 101
   Space Science and Applications, 46, 76
   space science budgets, 24–25
   Space Science Enterprise, 127
   Steering Committee, 9
National Radio Astronomy Observatory (NRAO), 17, 22
National Research Council, 5, 52, 106
National Science Foundation (NSF), 8, 23, 34
National Space Council, 106
Naugle, John E., 46, 52
Naval Research Laboratory (NRL), 9–11, 18, 23, 36, 44, 66–67
Neal, Valerie, 74

Near Infrared Camera and Multi-Object Spectrometer (NICMOS), 98, 130
nebulae, discovery of, 2, 27, 29, 41
Neel, Carr B., Jr., 16n15
Neugebauer, Gerry, 34, 37, 46
   Becklin-Neugebauer object, 27, 29, 41
   FIRST, 51–52
   infrared astronomy panel, 46, 68
   IRAS team listing, 62n5
   Mariner 2, 33
   quoted, 31
"New Start," 86, 94, 105
   for AXAF, 72–76, 86
   competition for, 72–78, 105
   for CGRO, 75
   defined, 71
   for Hubble, 75
   for SIRTF, xiv, 71–76, 83, 95–96, 123
   for SOFIA, 105
Newell, Homer, 35
Newton, George, 73, 109
Nixon, Richard, 36, 79

O

Obermann, Richard, 113
O'Brien, Brien, 74
observations
   conditions of, 22, 24, 38
   evaluations of, 48–49
   funding for, 53
   innovations, 71
   in orbit, 13–14, 70
   obstructions to, 13, 53, 69–70, 87–88, 102, 106–107
   tracking celestial objects, 13
   viewing window, 48
observatory programs, 13
observatories. *See individual names.*
Office of Management and Budget (OMB), 73–74, 77–81, 86
Office of Naval Research (ONR), 9, 14
Office of Scientific Research (U.S. Air Force), 15
Office of Space Science and Applications (OSSA), 46–47, **75**, 79–81
   Frank Martin at, 72
   funding for, 94
   Len Fisk at, 79, 94, 105
   Nancy Roman at (OSS), 11
optical astronomy, 11, 13, 20, 33. *See also* astronomy; infrared astronomy
   funding for, 53, 56
   influence of, 83
   and Kvant-1, 80
   and Space Shuttle, 46
Optical Astronomy Working Group, 43–45, 52
Optical Coating Laboratory, 31, 129n23
orbital mechanics, 102–105, 111, 114. *See also* orbits
Orbiting Astronomical Observatory (OAO) program, 14, 16, 34
Orbiting Geophysical Observatory (OGO) program, 14, 16
Orbiting Solar Observatory (OSO) program, 14–15, 30, 64n6
Orbiting Solar Observatory (OSO) model, **13**
Orbiting Solar Observatory (OSO-1), 15–16, **16**, 26, 63
Orbiting Solar Observatory (OSO-3), 16n15, 63
orbits. *See also* orbital mechanics
   polar versus equatorial, 70
   SIRTF performance in, 131–132, 136
   trailing-Earth, 102, 111–115, 137, 143, 151
Organization of the Petroleum Exporting Countries (OPEC), 46–47
Orion (constellation), 1–2, **2**, 5
   infrared observations of, 19, 68
   mythology of, **1**
Orion Nebula, 2–3, **2–4**
   aircraft-borne observations of, 25
   infrared spectrum of, 65–66
   protostar in, 27
   telescope observations of, 29
Ocampo, Cesar, 103

P

Paine, Thomas O., 35
particle accelerators, 15
passive cooled IRTs, 106–107, 111
Patel, Keyur, 134
Pellerin, Charles, 62, 138
   book, 137
   departure from NASA, 104
   Great Observatories program, 72–75, 77–81, 98, 125
   NASA Director of Astrophysics, 58, 72–76, 87, 90, 132
   preference for SIRTF at JPL, 92–93
   quoted, 148, 151
   SIRTF "New Start", 75–76
   SIRTF projected cost, 101–105
Pengra, Trish, 80
Penzias, Arno A., 17n17,
Perkin-Elmer Corporation, 53, 148
photoconductors, 6, **42**
physicists' contribution to astronomy, 14, 20–23, 25, 29–30, 37
Pioneer program, 10, 64, 83, 89–91
Pipher, Judith L., 25, 51, 117n2
planetary astronomy, 20–21, 83, **85**, 110, 118, 136
Planetary Society, the, 87
Pollock, James B., 65
Price, Stephan D., 30n3, 46
Principal Investigators (PIs), 20, 110, 139, **124**
   Shuttle-based IRT, 62
   SIRTF IRAC, 26, 62, 122
   SIRTF IRS, 25, 62, 122
   SIRTF MIPS, 26, 62, 122, 147
   SIRTF Phase B, 123

SIRTF redesign, 109
SIRTF spectroscopy, 97–98
public service, 9

Q

quasars, 14, 74
Quayle, Dan, 96

R

radiation, effects of, 14
radio astronomy, 10, 17, 22, 33
    funding for, 53, 56
radio waves (interstellar), 14, 16–17
Ramsey Committee, 9
RAND Corporation, 9
Rank, David M., 46, 49, 51, 68
Raven Industries, 15
Reagan, Ronald, 43n3, 56, 77–79, 86
    Soviet space program, 80–81
reclama, 79
Report of the Advisory Committee on the Future of the U.S. Space Program. See Augustine Report
Richards, Paul L., 51, 67–68
Ridgway, Stephen T., 51
Rieke, George H., 26–28, 49
    book, 137
    Design Optimization Study Team (DOST), 50
    FIRST, 51
    on Michael Werner, 68
    Multiband Imaging Photometer (MIPS), 122, 147
    personality of, 147
    quoted, 27–28
    Science Working Group PI, **69**, 70, 98, 110
    SIRTF cryostat, 130–131
    SIRTF facility scientist, 117
    SIRTF Multiband Imaging Photometer (MIPS), 62
    SIRTF project organization chart, **124**
Rieke, Marcia
    SIRTF chief advocate, 98–99, 101, 104, 138
    SIRTF cryostat, 130
    SIRTF/SOFIA advocacy program, 105
Rigel, 1, 3, 5
rocket-borne observations, 5, 8, 11, 13–14, 29
    costs of experiments, 25
    early, 29
    failures of, 23–24, 34
    scientists' preference for, 36
rockets, altitudes of, 8
rockoon, 8. *See also* astronomy, balloon-borne observations
Rockwell International, 52–53, 62
Roellig, Tom, 88n18, 117, 130, 150
Roman, Nancy, **6**, **13**, 52, 63
    and Hubble Space Telescope, 43
    NASA Director of Astrophysics, 9–11, 13–15, 23
    Optical Astronomy Working Group chair, 43, 45–46, 52, 63
    paper by, 10

    Payload Planning Working Group chair, 43
    personality of, 9–11
    quoted, 10–11, 14
    SIRTF Announcement of Opportunity, 54
Rosen, Robert, 93
Royal Observatory at Edinburgh, 106–108

S

Sagan, Carl, 21, 22n37, 87
Salyut, 35
satellites, **16**, 109
    early, 14
    gamma-ray experiment, 15n12
    infrared payloads, 13, 17–18
    orbiting observatories, 13–16, 26, 30
    SIRTF as, 45, 48, 54–55, 59, 75, 84, **113**
    Titan (moon), 6
Saturn, 20n30, 23, 87
Savedoff, Malcolm, 15
Schilling, Gerhardt, 11
Schmidt, Maarten, 14
science
    collaboration in, xi, 130–131, 137–151
    communication in, 147–151
    competition, lack of, 124, 143
    competition in, 56, 63–66, 87, 93–94, 105–110
    goals for SIRTF, 110
    government funding of, 8, 34, 53
    innovations in, 114–115, 125
    integrity in, 67–68, 76–77
    interdisciplinary, 22, 33, 62–63
    military contributions to, xii, 8, 17–18, 29–30
    national policy, 46
    objectives, 54, 57, 110
    parallel discovery, xii, 111
    partnerships in, 84, 138–151
    and politics, 53, 77, 138
    priorities for space program, 35–36
    as profession, 8–9, 11, 28
    as public service, 9
    rivalries in, 8–49, 62–66, 76–77, 109–111
    skepticism toward infrared, 21–22, 29, 31–33, 36, 65
Science Mission Directorate. *See* Office of Space Science and Applications (OSSA)
Science Working Group (SWG), xiii–xiv, 51, 53, 68–70, **69**, 83–84, 140
    advocacy for SIRTF, 98, 109
    dewar research, 125
    international collaboration, 88
    priorities of, 89–90, 110
    SIRTF cryostat, 130–131
    SIRTF Infrared Spectrograph (IRS), 130–131
    SIRTF orbit, 102–105
    SIRTF Phase B, 117, 123–124, 127
    SIRTF projected cost, 101–105
    SIRTF promotional brochure, 84
    SIRTF software, 134

SIRTF transfer to JPL, 95
"tiger team," 127
warm-launch concept, 111
working style of, 145
writings, 137
Scientific Advisory Board (U.S. Air Force), 9
Scientific Working Group. *See* Science Working Group (SWG)
Selzer, P.W., 37
semiconductors, 18–20
Senate Committee on Aeronautical and Space Sciences, 34, 52
sexism, 9–11
Share, Gerald, 15
Shuttle Infrared Telescope Science Accommodations Group (SIRTSAG), 49
Sidewinder missiles, 18
Simmons, Larry
  Jet Propulsion Laboratory (JPL), 130
  personality of, 109–110, 137–143, 145–146, 149–151
  quoted, 140–141, 144–147, 150–151
  SIRTF infrared spectrograph (IRS), 126
  SIRTF project manager, 100–101, 114, 123–124, 149–150
  SIRTF software, 134
SIRTF (Space Infrared Telescope Facility). *See* Space Infrared Telescope Facility (SIRTF); Spitzer Space Telescope
sky surveys,
  classified, 30n3
  infrared, 25–26, 29–33, 36, 43–44, 53
Smith, Bradford A., 34
Smoot, George, 17n17
social change, 11
SOFIA/SIRTF competition, 105–106, 108
Soifer, Baruch Thomas., 25, 27, 51, 110, 123, 130
  IRAS team listing, 62n5
solar astronomy, 36, 43, 54, 65
solar spectroscopy, 7–8
sounding rockets, 8
  altitudes of, 13
  and infrared telescopes, 23
South Atlantic Anomaly (SAA), 70
Soviet Academy of Sciences, 10
Soviet Union, 10, 35
  space program, 7, 10, 80–81
Space Infrared Telescope Facility (SIRTF), 70, **119–120**. *See also* Spitzer Space Telescope
  acronym change, 42
  acronym first use, 30, 50, 52
  actual costs, 125
  Ames oversight of, 91
  Announcement of Opportunity, xiv, 50, 58–59, 61, 69, 104, 117–118
  budgets for, 47, 54, 71, 87, 97–100, 124
  canceled, 98–99
  coloring book, 71
  competition with SOFIA, 105–106, 108
  construction of, 27–28, 117–131, **119–122**, 124–125
  cost, actual, 134
  Critical Design Review, 134
  cryogenic telescope assembly, 123
  cryostat, **89**, 126–131
  delayed, 58, 95, 97, 134
  design (launch-ready), 123
  Design Optimization Study Team (DOST), 53
  design (preliminary) 41–42, **41**, 44–45, 53, 118n5
  design summary document, 53
  dewar, **89**, 106, 129–131
  dewar pressurization, 125–127
  dimensions, xi, 111, 119
  Edison/SIRTF competition, 107–108
  effects of radiation on, 70
  as facility (opposed to telescope), 30
  fate of, 136
  fault protection, 132
  feasibility study of, 51
  in federal budget, 113
  filter delamination, 125
  flight-ready design (Phase B), 118
  foundations for, 20–28, 29
  as free flying satellite, 45, 48, 54–55, 59, 75, 84, **113**
  Great Observatories program, 72–81, 84
  Infrared Array Camera (IRAC), 26, 62, 70, 92, 118–119, **121–122**
  Infrared Array Camera (IRAC), successes of, 136
  infrared detectors, 70–71, 117
  Infrared Processing and Analysis Center, 123
  Infrared Spectrograph (IRS), 61–62, **121–122**, 123, 125–127, 129–131
  instruments, 41–42, 118–119, 123, 131–132
  integration flow system, **89**
  internal reviews, 123
  IRAS program, 112
  JPL Programmatic Commitment Agreement, 114
  launch, xi, 134–135
  management bidding outcome, 94
  Management Center, 105
  mirror, 112, 118–119, **121**
  mission operations, 134
  Multiband Imaging Photometer (MIPS), xi, 26, 62, 118–119, **120–122**, 123
  Multiple Instrument Chamber (MIC), 119, **121**, 126, 129–131
  and national policy, 46
  "New Start," xiv, 71–76, 95–96, 105–106, 123
  Non-Advocate Review (NAR), 123
  "non-competition competition," 92
  objectives, 54, 118, 123
  observation window, 70, 104
  orbit, performance in, 131–132, 136
  orbit, trailing-Earth, 102, 111–112, 114–115, 136n61, 137
  orbit of, 69–70, 84, 87–88, 102–105

Phase B, 95–96, 113–114, 123–127, 129–135
Phase C, 95, 113–114, 123–124
Phase D, 95, 114, 124
Phase E, 95, 114, 117
planetary astronomy, 83
pointing control system (PCS), 132
Preliminary Design Review (PDR), 123
priority of, 66, 73–74, 83, 87, 90, 97
project management bids, 92–94, 132
project organization chart, **96**, **124**
project schedule, 71–81, **71**
projected cost of, 101–112
prototypes, 118
publications on, 73–74, 81, 84, 137
redesigned, 108–112
renamed Spitzer Space Telescope, xi, 34, 42, 136
report timeline, 52
rocket launch options, 102–104
safe mode, 132
scaled back, 88, 98–100, 104–105, 109–114, **112**, 118–119, 123
Science Accommodations Group, 52
Science Center, 92, 123, 134
science goals of, 110
Science Working Group, 53
as Shuttle-based facility, 41, **44**, 47–51
Shuttle-based facility integration report, 53
SIRTF/SOFIA review, 106
software, 124–126, 131–135
specifications, **42**
spectroscopy, 131–132
Spitzer Science Center, 27
study team, 54
successes of, 135–136, 147
SWG planning retreat, 109
system integration flow, **89**
testing of, 130–131
"tiger team," 127
timeline, 47, **71**
transfer to JPL, 94–97, 102, 105
Über system-design team, 144
warm-launch concept, 111–114, 117
space program (U.S.), 14, 21, 33
 budgets for, 77–78
 criticism of, 34–36, 56, 79, 84–86
 flight testing, 91
 future of, 96–97
 hardware, 62
 infrared telescopes (IRT), 45
 and policy makers, 35–36
 priorities of, 53
Space Science Board (SSB), 8–9, 33, 52–53. *See also* Committee on Space Astronomy and Astrophysics (CSAA)
 infrared recommendations, 17–20, 46
 priorities of, 21n35
Space Shuttle (U.S.), 21, 35–36
 budgets for, 47
 Challenger, xiii, 43n3, 78–81, 85n6, 99
 Columbia Accident Investigation Board, 101n1
 contamination by, 47–50, 55
 dimensions, 44
 engine bells, 107
 engineering studies, 48–49
 observations (shuttle-based), 30
 Payload Planning Working Groups (final report), 52
 payloads, 29–30, 42–46, 58, 85n6, 111
 Payloads: Committee on Aeronautical and Space Sciences (hearing), 52
 politics of, 46–47
 priority of, 89
 problems with, 84–86, 101
 scientific uses of, 43
 servicing AXAF, 88
 servicing Hubble, 100
 Shuttle accommodations/mission system engineering study, 52
 Shuttle Infrared Telescope Science Accommodation Group (SIRTSAG), 49
 as SIRTF mount, 44–45, 50, 53, 55, 57–58
 SIRTF sortie mission, 38, 42
 Space Shuttle Sortie Workshop (proceedings), 52
space-borne observations, 33–36, 35, 136
Spacelab, 2, 50, 53, 58, 62, 69
spectra, 4–5, 11, 13–14
 infrared, 19, 62
spectroscopy, 6–8, 38, 110
 defined, 45
 stars, 21
Spehalski, Richard, 96, **96**, 99
Spitzer, Lyman, Jr., 33–34, 97, 136
 quoted, xi
Spitzer Science Center, 27
Spitzer Space Telescope, xi–xiii, 34, 42, 134, 136–137. *See also* Space Infrared Telescope Facility (SIRTF)
Sputnik program, 7, 10, 15n12, 91n26
Stanford University, 37, 66
starlight, 3–5, 10, 23, 122
 protostars, 19, 27
Stecher, Theodore P., 43
Stein, Wayne A., 51
Steward Observatory, 44, 51
Straetker, John, 134, 138
Strategic Defense Initiative (SDI), 78, 81
Stratospheric Observatory for Infrared Astronomy (SOFIA), 105–106, 108
Street, J.C. (Jabez Curry), 26
Sunyaev, Rashid Alievich, 80

T

Tananbaum, Harvey, 74n33, 98
theoretical limit, 19–22
Thompson, Rodger I., 51
Thronson, Harley, 106–108

Titan, infrared observations of, 23
Titan (moon), 6
Titan rocket, 102, 109, 112, **113**
Tolivar, Fernando, 134
Topography Experiment (TOPEX), 77n45
Total Quality Management, 101
Townes, Charles, 67
trailing-Earth orbit, 102, 111–115, 137, 143, 151
Trapezium constellation, 2n2, 3
Truly, Richard, 101
TRW Space and Technology Group, 101
200-inch Palomar telescope (Hale), 9, 14, 19, 27, 34
Two-Micron Sky Survey, 31, 33
    Two-Micron Sky Survey telescope, 31, **32**

U

ultraviolet astronomy, 8, 11, 13, 17–18, 54
    funding for, 56
universe
    age of, xii, 17, 110, 118
    origin of, 68, 136
    size of, 110
University College London, 46
University of Arizona, 63
    bolometers, 20
    Dale Cruikshank at 83
    Peter Eisenhardt at, 95
    FIRST, 51
    infrared astronomy panel, 46
    William Hoffmann at, 43, 46
    Douglas Kleinmann at, 27
    Gerard Kuiper at, 21, 26, 83
    Frank Low at, 21–23, 27, 37, 46
    Lunar and Planetary laboratory (LPL), 21
    George Rieke at, 26, 46, 98, 122, 147
    SIRTF meetings, 139
    SIRTF project organization chart, **124**
University of California–Berkeley, 36, 67
University of California–San Diego, 36
University of California–Santa Cruz, 68
University of Chicago, 5, 9, 22n37, 51
University of Hawai'i, 51, 62, 66
University of Iowa, 9
University of Minnesota, 20n30, 36, 43, 49, 51
University of Rochester, 15–16, 26, 51
University of Texas, 20–21, 103
University of Texas at Austin, 51, 106–107
University of Wyoming, 49, 51, 106–108
U.S. Air Force, 9, 15, 30, 135
    Cambridge Research Laboratories (AFCRL), 30, 36, 38, 46, 49
    infrared surveys, 30n3, 57
U.S. Army, 7, 10
U.S. Navy, 18, 23, 36. *See also* Cambridge Research Laboratories (AFCRL); military contributions to science.
    China Lake facility, 18
    Naval Research Laboratory (NRL), 9–11, 18, 23, 36, 44, 66–67

V

V-2 rocket, 7–8, 34
Van Allen, James, 7–9, 14
Van Allen radiation belts, 14, 70
van de Kamp, Peter, 11
Vanlandingham, Don, 126
Venus, 19, 33, 65
von Braun, Wernher, 8
Voyager program, 83

W

Walker, Russell G., 30n3, 36, 49, 62n5
warm-launch concept, 111–114, 117
Washburn Observatory 44
Wdowiak, Thomas, 43
Webb, James, 21n35, 35
Webb Space Telescope (JWST), 108n24, 111n33, 122
Weedman, Daniel, 104–106, 113, 138
Weiler, Edward J., 127, 134
Werner, Michael, 25, 146
    advocacy for SIRTF, 108
    article by, 62n5, 123
    book, 137
    FIRST, 51
    influence of, 68
    masers, 67
    personality of, 110, 117, 130–131, 137–141, 143
    quoted, 139, 145–146, 147, 150
    Science Working Group, 68–71, 103
    SIRTF cryostat, 130–131
    SIRTF fault protection, 132
    SIRTF Infrared Spectrograph (IRS), 127, 129–131
    SIRTF Phase B, 117
    SIRTF project scientist, 63, 66–71, 95, **96**
    SIRTF redesigned, 113
    SIRTF software, 134–135
Westphal, James, 18–19, 33
White Sands Missile Range (Proving Ground), 7–8
Wide Field/Planetary Camera Number 2 (WFPC2), 100
Wildey, Robert, 33
Wilson, Robert K., 17n17, 134–135
Witteborn, Fred C., 51–52, 68–71
    Ames astrophysics chief, 36–38, 41–42, 46
    FIRST, 51
    Greenstein report, 58–59
    Hughes report, 50, 52
    influence of, 63–66
    infrared astronomy panel, 46
    infrared telescopes (IRT) observation window, 48
    quoted, 37–38, 42–43, 48–49, 54
    Science Working Group, **69**
    SIRTF Announcement of Opportunity, 58–59
    SIRTF conceived by, 36, 50, 138
    SIRTF deputy project scientist, 63–66
    SIRTF spectrometer proposal, 62
    SIRTF study team, 54

Space Infrared Telescope Facility (SIRTF) first presentation, 52
Very early system design study (SIRTF), 52
Working Group on Infrared Astronomy, 43
Working Group on Infrared Astronomy, 43
Working Group on Optical Astronomy, 33
Working Group on Radio and Radar Astronomy, 33
Wright, Edward L., 27, 51, 63, 69

X

x-ray astronomy, 11, 36, 43, 54
    Einstein Observatory, 92
    funding for, 56
    Kvant-1, 80
x-rays, 13, 17–18

Y

Yerkes Observatory, 5–6, 8–9, 11, 44, 49
Young, Lou, 41
    FIRST, 51
    Hughes report, 50, 52
    Mobius strip SIRTF schedule, **71**
    SIRTF project manager, 63
    SIRTF study team, 54
    Space Infrared Telescope Facility (SIRTF) first presentation, 51n27, 52
    Very early system design study (SIRTF), 52

Z

zero gravity, 8, 37

www.ingramcontent.com/pod-product-compliance
Lightning Source LLC
Chambersburg PA
CBHW082203220526
45470CB00010B/3026